VoIP SERVICE QUALITY

McGRAW-HILL NETWORKING AND TELECOMMUNICATIONS

Demystified

Harte/Levine/Kikta	*3G Wireless Demystified*
LaRocca	*802.11 Demystified*
Muller	*Bluetooth Demystified*
Taylor	*DVD Demystified*
Hoffman	*GPRS Demystified*
Camarillo	*SIP Demystified*
Shepard	*SONET/SDH Demystified*
Topic	*Streaming Media Demystified*
Symes	*Video Compression Demystified*

Developer Guides

Guthery	*Mobile Application Development with SMS*
Richard	*Service and Device Discovery*

Network Engineering

Rohde/Whitaker	*Communications Receivers,* 3/e
Sayre	*Complete Wireless Design*
OSA	*Fiber Optics Handbook*
Lee	*Mobile Cellular Telecommunications,* 2/e
Bates	*Optimizing Voice in ATM/IP Mobile Networks*
Roddy	*Satellite Communications,* 3/e
Simon	*Spread Spectrum Communications Handbook*
Snyder	*Wireless Telecommunications Networking with ANSI-41,* 2/e

Professional Networking

Smith/Collins	*3G Wireless Networks*
Collins	*Carrier Grade Voice over IP,* 2/e
Minoli	*Enhanced SONET Metro Area Networks*
Minoli/Johnson/Minoli	*Ethernet-Based Metro Area Networks*
Benner	*Fibre Channel for SANs*
Bates	*GPRS*
Minoli	*Hotspot Networks*
Bates	*Optical Switching and Networking Handbook*
Wang	*Packet Broadband Network Handbook*
Sulkin	*PBX Systems for IP Telephony*
Russell	*Signaling System #7,* 4/e
Saperia	*SNMP on the Edge*
Ohrtman	*Softswitch*
Nagar	*Telecom Service Rollouts*
Hardy	*VoIP Service Quality*
Karim/Sarraf	*W-CDMA and cdma2000 for 3G Mobile Networks*
Bates	*Wireless Broadband Handbook*
Faigen	*Wireless Data for the Enterprise*

Security

Hershey	*Cryptography Demystified*
Buchanan	*Disaster Proofing Information Systems*
Nichols	*Wireless Security*

VoIP Service Quality

Measuring and Evaluating Packet-Switched Voice

William C. Hardy

McGraw-Hill
New York • Chicago • San Francisco • Lisbon
London • Madrid • Mexico City • Milan • New Delhi
San Juan • Seoul • Singapore
Sydney • Toronto

The McGraw-Hill Companies

Library of Congress Cataloging-in-Publication Data

Hardy, William C.
VoIP service quality : measuring and evaluating packet-switched voice / William C. Hardy.
p. cm.
Includes index.
ISBN 0-07-141076-7 (alk. paper)
1. Internet telephony. I. Title.

TK5105.8865.H37 2003
004.6—dc21 2002044384

1 2 3 4 5 6 7 8 9 0 DOC/DOC 0 9 8 7 6 5 4 3

ISBN 0-07-141076-7

The sponsoring editor for this book was Marjorie Spencer, the editing supervisor was Stephen M. Smith, and the production supervisor was Sherri Souffrance. It was set in Century Schoolbook by Deirdre Sheean of McGraw-Hill Professional's Hightstown, N.J., composition unit.

Printed and bound by RR Donnelley.

This book is printed on recycled, acid-free paper containing a minimum of 50% recycled, de-inked fiber.

This, too, is for Adriana, with love.

CONTENTS

PREFACE

The focus of this book is the narrow question of how to assess quality of packet-switched voice services in general and VoIP services in particular. The approach taken in answering this vexing question is one that I have exploited to very good effect in more than 35 years' working in the general area of test and evaluation of telecommunications systems. In applying this technique I

- Imagine myself using the system that is the subject of evaluation
- Decide what I would be concerned about if I were to be its user
- Research the technology of the system to the extent necessary to understand the mechanisms determining system performance that affected what I would experience with respect to those concerns
- Formalize the relationships between system performance and user perception of quality gleaned in this manner

The result is invariably a system of measurement and evaluation whose rationale is almost self-evident to even the most casual student of the system and often smacks of trivial observation to persons immersed in its intricate, microscopic technical details. The present treatment of packet-switched voice services is probably no exception. What is presented here will to some be painfully long on development of general measurement concepts and measurement technology and short on the specific details of implementation of the measures and models defined. As a consequence, the reader should not expect, for example, to find in this volume a complete set of equations for calculating PESQ (Perceptual Evaluation of Speech Quality) measures. What the reader should walk away with, however, is a very good understanding of the basis for PESQ, how it was developed, its strengths and weaknesses for various applications, when to use it, when to avoid its use, and, most important, why. The objective is to arm the reader with the perspectives and understanding that will enable a similar assessment of the next new be-all, end-all technique for predicting likely user assessment of quality of the next new packet-switched voice service, and the ones after that, and the ones after those.

—William C. Hardy

INTRODUCTION

In today's environment nearly all end-to-end telephone connections are set up via *circuit switching,* whereby node-to-node links in an origin/destination connection are set up via interconnects, and the connection is maintained exclusively for exchanges of information between the origin and destination until it is torn down. An alternate way of setting up end-to-end connections that is widely used for transmission of data is *packet switching,* such as that used in the Internet, whereby origin-to-destination connections are effected by node-to-node, store-and-forward relay of small segments of data sets that are reassembled at the destination.

Since digital data sets transmitted across a packet-switched network might as easily comprise digitized voice signals as anything else, there is no issue as to whether voice can be transmitted via packet-switched network. However, the essential question remains as to whether, and/or under what circumstances, packet-switched transport will adequately support telephony and other applications, such as multimedia conferencing, requiring near-real-time, multidirectional exchanges of voice signals.

The possibility of creating such interactive packet-switched voice services creates both opportunities and a problem for development. The development of viable packet-switched voice transport creates opportunities both for merging the transport of voice and data services, thereby realizing substantial operational flexibility and economies in switching voice service, and for development of new services, such as integrated messaging, that would exploit the characteristics of a packet-switched network. The problem is that it is not clear whether, or under what circumstances, the quality of packet-switched voice services will be satisfactory for their intended uses.

To resolve this quandary and safely exploit packet-switching technology where possible, communications service managers must be able to assess the operational characteristics of packet-switched voice services relative to the needs of their application and determine how users are likely to perceive the quality of those services. At the same time, telecommunications service providers must be able to configure and operate packet-switched networks in a way that assures requirements for user perception of quality of service (QoS) are met.

The material in this book is intended to facilitate the development of capabilities for accomplishing these ends by setting forth a framework for measurement and evaluation of *perceived quality of service* of packet-switched voice services relative to different applications. It is based on the more general foundations for measurement and evaluation of telecommunications QoS presented in Ref. 1, often appealing to concepts introduced in that book and adding the specifics needed for their application to packet-switched voice services.

The presentation is divided into three parts:

- Part 1, Foundations, contains all of the background material needed for understanding the factors that affect users' perception of, and satisfaction with, quality of packet-switched voice services. It covers the basic notions of quality of service derived from analysis of user concerns with quality, together with descriptions of the system-level interactions that determine what users will experience when voice exchanges are packet-switched.
- Part 2, Measurement and Evaluation of Voice Quality, turns to the central question of ways and means of gauging likely user perception of the quality of packet-switched voice services with respect to the audible quality of voice and naturalness of the exchanges. It describes commonly used techniques for measuring and analyzing voice quality, together with procedures for using such measures to determine what levels of performance of the packet-switched transport are needed to ensure that the voice quality will be acceptable to users.
- Part 3, Other Aspects of Quality of Service, concludes this book with brief descriptions of the ways and means of measuring and gauging likely user perception of packet-switched voice services with respect to the other user concerns with telecommunications QoS described in Ref. 1 and some of the unique quality requirements associated with some kinds of packet-switched voice services.

VoIP SERVICE QUALITY

PART 1

Foundations

The object of study of this book is *perceived quality of packet-switched voice services.* The purpose is to describe and suggest applications for techniques by which objectively measured characteristics of those packet-switched voice services can be analyzed to predict user satisfaction.

Even the object of study can be described unambiguously only with the assistance of detailed definitions and distinctions, and the viability of different evaluative concepts and models can be appreciated only in light of basic understanding of packet-switched voice systems. Accordingly, we begin here with a presentation of fundamentals, covering such basics as notions of voice services, measurement and evaluation of quality of voice service, and differences in implementation and performance between packet- and circuit-switched voice services. Although these topics may look familiar to the knowledgeable reader, it is important for everyone to become familiar with this part of the book. Because the foundations laid here are essential perspectives, rather than recapitulation of conventional material, on these topics, much of what is presented later in the book may look like *jabberwocky* absent the assistance of the definitions and concepts given here.

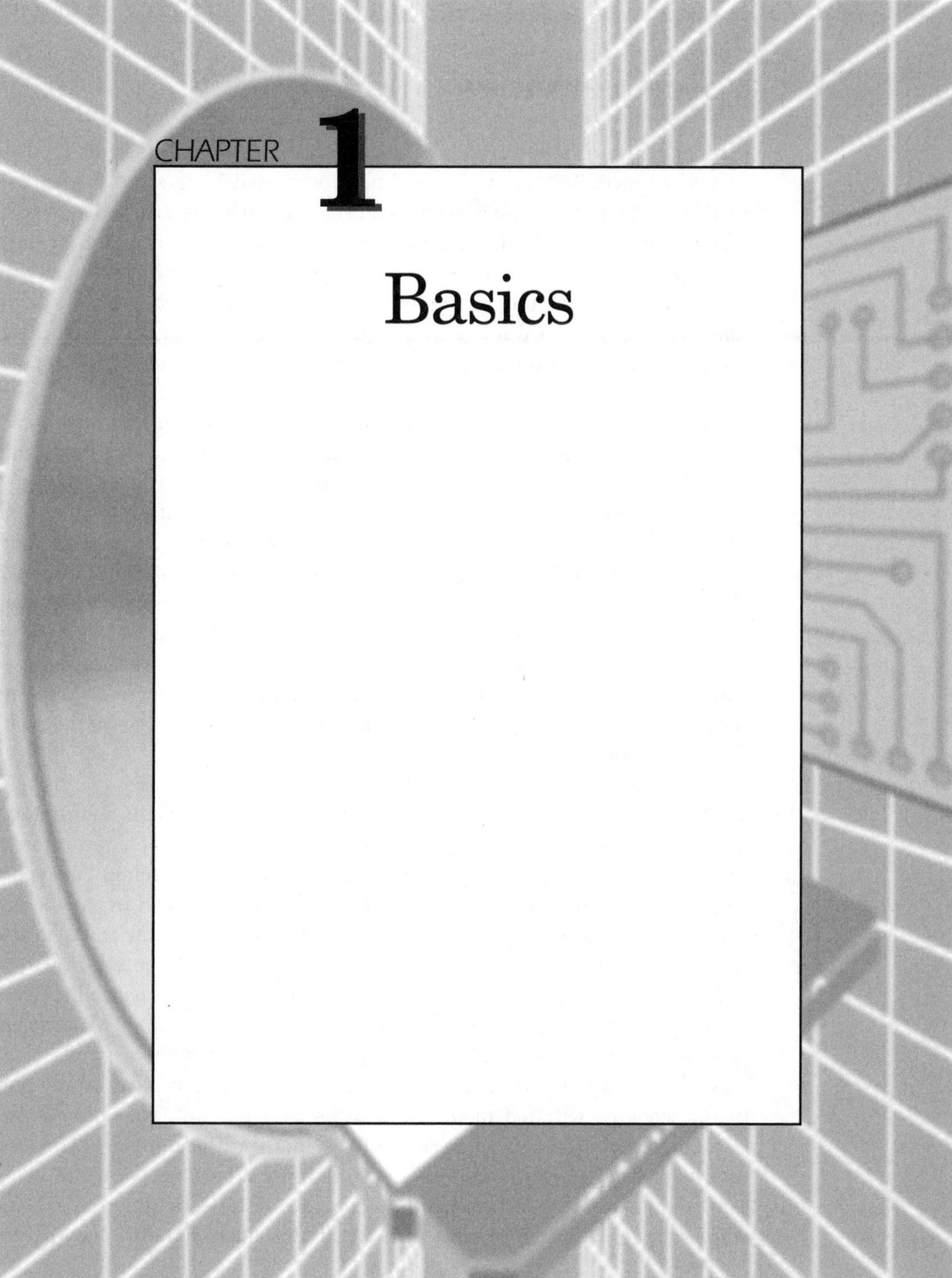

CHAPTER 1

Basics

Voice Services

It must be understood from the outset that although this book focuses on packet-switched voice, we are not concerned here simply with the ability to transmit voice signals across a packet-switched network without unacceptable deterioration of voice quality. Since digital data sets transmitted across a packet-switched network might just as easily comprise digitized voice signals as anything else, there is no question that even very high fidelity digitized voice signals can be transmitted across a packet-switched transmission network with negligible loss of fidelity.

Rather, we are concerned with the ability to digitize and transmit voice signals across a packet-switched network and the ability to do this in a way that supports near-real-time, multidirectional voice exchanges. To distinguish this application, transmission capabilities designed to support such interactive exchanges of voice are referred to here as *voice services.* Under this convention, for example, the ability to transmit a digitized recording of a voice message via a streaming voice system does not constitute a voice service, because the transmission is not effected in near real time. Similarly, even the unbuffered, direct transmission of voice as part of a video clip fails to qualify as a voice service because no accommodations of the kind of interactive exchanges that would occur over a picture telephone are required.

Such voice services are often described in technical discourse as Vo*X,* where Vo stands for *voice over* and *X* represents the transmission protocol used in the host packet-switched network. Thus, for example, an interactive voice exchange capability carried over packet-switched transport employing the Internet protocol (IP) is frequently described in the technical literature as VoIP. This nomenclature conveys information as to the type of network in which the voice service is to be implemented. However, it does not convey any information as to the kind of voice service involved. Consequently, the use of the Vo*X* (e.g., VoIP, VoFrame, VoATM) descriptors sometimes fosters the erroneous notion that there is a single voice service contemplated or implemented in each medium. In fact, in any particular packet-switched medium, such as the Internet, we may see the implementation of a wide variety of distinctly different voice services, each with its own requirements and functions. Where necessary to avoid confusion, lowercase letters will be added after the *X* to denote a specific voice service. Thus, for example, later in this book you will see *VoIPtpt* used to distinguish general-use voice transport via IP networks from the more special case of on-net telephony, denoted *VoIPtel.*

Quality of Service (QoS)

The other ambiguity in descriptions of packet-switched voice services that must be clarified at the outset is what is meant by *quality of service.* There are at least three distinctly different referents for the term QoS that appear in technical discourse on the subject.

1. *Capabilities for, or the classes defined to achieve, preferential handling of different types of traffic in packet-switched networks.* In much of the data networking literature, particularly that dealing with the Internet protocol, the term QoS is understood to mean a preferential class of service to which a particular transmission may be assigned. The class is created by specification of particular handling or routing capabilities that can be employed to afford specified types of traffic priority use of the available bandwidth. Thus, for example, Ref. 2, p. 189, describes QoS as follows:

> In this book, QoS refers to both class of service (CoS) and type of service (ToS). The basic goal of CoS and ToS is to achieve the bandwidth and latency needed for a particular application. A CoS enables a network administrator to group different packet flows, each having distinct latency and bandwidth requirements. A ToS is a field in an Internet Protocol (IP) header that enables CoS...

2. *Intrinsic quality of service.* When traffic is carried via a packet-switched network, with or without application of QoS capabilities, the handling of the traffic will achieve certain operational performance levels under various levels of demand. Those characteristics that can be measured by the provider without reference to user perception of quality but that will, nonetheless, affect user perception of quality are referred to in Ref. 1 as defining *intrinsic* QoS. It is generally agreed that for packet-switched services such intrinsic QoS is characterized by

- *Latency.* The time it takes a packet to get across the packet-switched network to its destination
- *Jitter.* The variability in packet latency
- *Dropped packet rate.* The frequency with which packets do not get to their destination in time to be used

For any class of traffic these characteristics will, in general, depend on the size of the demand and the amount of bandwidth allocated to that traffic.

3. *Perceived quality of service.* Perceived quality of service is distinguished from intrinsic QoS as being what results when the service is actually used. Perceived QoS is, then, determined by what users experience as the effects of intrinsic QoS on their communications activities, in their environment, in handling their demand, and how they react to that experience in light of their personal expectations. It is perceived, rather than intrinsic, QoS that ultimately determines whether a user will be satisfied with the service delivered.

Objectives of Measurement and Evaluation

Notice, then, that if we fail to distinguish between the variety of commonly understood meanings of the term QoS, we might assert, without fear of contradiction, that

> Without QoS, the QoS for most packet-switched networks will not support adequate QoS.

To make sense of this sentence, we need to use the more precise terms introduced in the previous section:

> Without preferential QoS, the intrinsic QoS for most packet-switched networks will not support adequate perceived QoS.

This sentence now asserts that our objective here is to detail ways and means of determining levels of intrinsic QoS for packet-switched voice services that will assure adequate perceived QoS when those services are fielded. In doing this, it is necessary to

- Describe measures of perceived QoS that can be readily quantified to reliably gauge likely user satisfaction with various packet-switched voice services.
- Relate those measures of perceived QoS to measures of intrinsic QoS to create a basis for determining the characteristics that must be achieved in the packet-switched network to assure that perceived QoS is acceptable.

Principal User Concerns

For myriad reasons that will not be elaborated here, the point of departure for the first step of defining measures of perceived QoS recommended in Ref. 1 is a description of likely *user concerns* regarding QoS. Such concerns are fostered by users' experiences with less than satisfactory quality on similar services and are usually expressed as doubts or questions seeking positive reassurance. For the case of packet-switched voice, users' principal concerns are with the *connection quality,* i.e., the quality of conversations carried over the service, as typified by concerns with the quality of what is heard:

- Will connections exhibit impairments that will make it difficult to hear and understand what is being said? Will I be bothered with echo when I try to talk?
- Will the distant speakers' voices sound natural? Will I be able to readily recognize different speakers?

and connection usability:

- Will the natural conversational rhythms and intonations be preserved in the flow of speech between me and the distant speakers?
- Will the service support natural conversational rhythms and speech patterns in interactive exchanges of information?

In expressing concerns like these, the prospective users of a new voice service will necessarily be synthesizing, or reacting to, their previous experiences using similar voice services. Thus, the concerns with voice quality will focus on familiar impairments experienced on telephone calls completed via other voice services. Similarly, users who have experienced, and been irritated by, the kinds of delays that occur in international long-distance calls completed via satellites will express concerns with connection usability by asking whether packet switching can result in similar delays.

The other, universal concerns regarding QoS of a telecommunications service identified in Ref. 1 are listed in Table 1-1. As described in Part 3, the transition to a packet-switched network will create differences in performance that may have deleterious effects on user perception of quality with respect to some of these. However, none of those concerns looms nearly as large as the widespread concern as to what packet switching will do to the quality of voice services.

TABLE 1-1

User Concerns with Quality of Telephone Services Other Than Voice Quality

Accessibility	Will I be able to get to the service when I want to use it? How long will I have to wait if I can't? How often will the wait be really bothersome?
Routing speed	How long does it take before I know that a connection is being set up? Is the time predictable?
Connection reliability	When I dial a number, will the service set up a connection to the distant station or let me know when the station is busy?
Routing reliability	If I dial the number correctly, will the service set up the right connection?
Connection continuity	Will my voice connection stay up until I hang up? Will data exchanges complete without premature disconnection?
Disconnection reliability	Will the connection be taken down as soon as I hang up? What happens if it isn't? Is there someone who will believe me when I tell them that I did not talk to my mother-in-law for six solid hours, and correct the billing?

Applications

The principal thrust of this book, then, is to examine such user concerns to develop measures of perceived QoS and then to clearly correlate those measures with the classical intrinsic measures of QoS for packet-switched voice services. The machinery thus developed is expected to greatly facilitate resolution of numerous critical issues with respect to packet-switched voice services that require assessment of likely user perception of voice quality, such as:

- What levels of packet latency, jitter, and dropped frame rate should I design to for different kinds of services, and to provide acceptable quality without paying more than I need to?
- Provider A is offering me a service with intrinsic quality specifications S_A for $X, while provider B is offering different quality specifications S_B for $Y. Which represents the better deal? Will either service actually satisfy my users?
- Will a packet-switched voice application work for this particular kind of service?
- *How do I know what to tell people to stop all these questions?*

CHAPTER 2

Principal System-Level Determinants of Connection Quality

Cogent answers to the questions raised at the end of Chap. 1 will depend on values of the measures of intrinsic QoS—packet latency, jitter, and dropped frame rate—and the way that those performance characteristics affect the QoS manifested to users. In particular, as will be described in this chapter, relationships between measures of intrinsic and perceived QoS with respect to voice connection quality will be determined by three characteristics of the system implementing the packet-switched voice service:

1. *Voice codec (coder/decoder)*, which determines how the voice signals are digitized for transmission
2. *Packetization scheme*, which sets the duration of the segments of digitized voice payload transmitted in each packet and the size, in number of bits, of packet headers
3. *Size of the jitter buffer*, which determines codec resiliency to variations in packet transmission delays

Voice Codecs

Voice codecs (see App. A) are designed to International Telecommunication Union (ITU) standards, which specify how segments of analog voice signals are to be encoded into digital data streams. The design of the particular codec used to digitize voice signals carried via a packet-switched network determines both the minimum number of bytes that can be reasonably included in a voice packet and the throughput of packet bits that must be achieved in order to transmit a digitized voice signal. Table 2-1 shows, for example, the characteristics of three of the codecs that are most widely considered for possible use in setting up VoIP services. All three are based on an 8000-hertz (Hz) sampling rate for analog voice signals. However, as shown in the table, the differences in encoding techniques create substantial differences in both the minimum duration of the segment of voice that is sampled and the amount of data transmitted to support regeneration of the analog signals at the distant end. The codec characteristics shown in the table, then, directly affect two characteristics of the voice signals heard by users over a digitized voice connection: delays and signal fidelity.

Delays In order to model the voice segments with the duration shown in Table 2-1, the codec must have the complete segment and possibly

TABLE 2-1

Comparison of Characteristics of Voice Codecs

Codec	Encoding Technique	Voice Segment Duration, ms	Encoded Segment Size, bits	Data Rate, bit/s
G.711	Pulse-code modulation (PCM)	0.125	8	64,000
G.723.1	Multipulse maximum-likelihood quantization (MP-MLQ)	30	189	6,300
G.723.1	Algebraic-code-excited linear prediction (ACELP)	30	158	5,300
G.729	Code-excited linear prediction (CELP)	10	80	8,000

more available for processing. For example, the G.723.1 codec must receive and buffer 37.5 milliseconds (ms) of digital voice samples before the encoding with the numbers of bits shown can be effected. Use of the G.723.1 codec therefore increases connection characteristics like echo path delay and round-trip conversational delays by at least 67.5 ms (= 37.5-ms encoding time and 30-ms decoding time) over the continuous transmission of signals encoded with the G.711 codec in today's circuit-switched voice services.

Signal Fidelity The digitization of voice approximates the continuous electrical signal representing the acoustic waveforms that excited the microphone, and the effects of those approximations result in deformations of the electrical signals intended to excite the telephone earpiece. Consequently, even when digital transmission is perfect, there will invariably be differences between the injected analog waveform and that extracted at the distant end. For all standard codecs, the differences between the injected and extracted analog waveforms given error-free digital transmission are not expected to be great enough to materially affect the quality of voice transmission with respect to intelligibility or speaker recognition. The waveform distortions produced by digital transmission with a particular codec may, however, be great enough to produce a noticeable degradation of what users describe as the "clarity" of the voice transmissions. Moreover, deviations from expected signal characteristics and digital transport error rates will begin to produce waveform deformations that are clearly manifested to users as "speech distortion" as described for test subjects participating in subjective tests of voice quality. For example, with very high signal levels and low line noise levels at the extraction side of a voice transmission digitized with a PCM codec, the quantizing noise will become perceptible, causing users to begin to report that the speech is distorted, because the person's voice sounds unnaturally "raspy." Similarly, high signal levels on the injection side of a PCM codec will result in unnatural amplitude clipping of the extracted waveform that produces an unnatural sounding voice. And, for all codecs, bit errors in transmission will deform the digitized approximation of the injected waveform in ways that may be noticeable to users as speech distortion, depending on the bit error rate and the specific encoding algorithms used in the codec.

Other, optional features in the way that a codec is implemented may directly affect user perception of the incidence or severity of recognizable impairments. These include silence suppression, comfort noise, and packet loss concealment.

Silence Suppression For purposes of minimizing the data throughput needed to transmit digitized voice signals, codecs may be configured to monitor the injected analog signal and digitize and transmit only what appears to be voice. This optional feature in a codec is variously referred to as voice activity compression (VAC) or voice activity detection (VAD) and silence suppression. In addition to reducing the throughput required to support a voice service, the use of silence suppression has the salubrious effect of reducing the perception of incidence and severity of "noise." Against these good effects, however, silence suppression can have two deleterious effects on user perception of voice quality. The first is that in low-volume speech signals the VAD will be slow in detecting soft beginnings and endings of words and syllables, producing what is known as *VAC clipping*, under which users notice that expected sounds are missing from the received speech. Such clipping can become a major irritant when a user is trying to maintain a conversation. The second effect is that silence suppression produces a complete absence of signal on the line at the distant end. For users of the circuit-switched telephony who are accustomed to at least some line noise on a connection, this can result in a disconcerting misperception that the line has gone dead. When it occurs, such confusion prompts users to rate the call as "difficult" or "irritating," no matter how good it is otherwise.

Comfort Noise One of the sometimes disconcerting characteristics of all digital voice connections is that there is absolutely no noise on the line when no one is talking. Since users commonly experience low levels of noise when any part of the connection is analog, this "deep null" condition makes it appear that the line has gone dead. To circumvent this problem, the decoder may be programmed to insert a low-level pseudorandom noise signal whenever there is no signal being received. Such inserted noise is referred to as *comfort noise*, because it reduces the incidence of the perception of deep nulls as dead lines. At the same time, it creates an opportunity for the system to generate something that will increase the user's perception of incidence and severity of noise. The perceived quality of connections will, therefore, be affected by the choice to insert comfort noise and the procedures for doing so.

Packet Loss Concealment For codecs that are used in packet-switched voice services, there is a possibility that some frames of the digitized voice signal will not arrive by the time they are needed to regenerate the next voice segment. To create resiliency to the effects of such missing samples on the waveforms reconstructed at the distant end, the encoder may be programmed to fill gaps in sampled data. Typical devices of this kind include simple repetition of the last frame received or generation of an artificial voice segment consonant with immediately previous frames of sampled data. Such compensation for dropped frames can substantially reduce the deleterious effects of packet loss on user perception of the incidence and severity of speech distortion.

Packetization

Another system characteristic that will greatly affect the way that packet loss across a packet-switched network will affect the user perception of speech distortion over a connection is the way that the packets are constructed for transmission. A packet-switching protocol is implemented by gathering a set of bits to be transmitted and adding the information needed for routing and handling those bits across the network. The added bits are referred to as the *packet header* or, more colloquially, the *envelope,* and the injected bits to be delivered are referred to as the *payload.* Because the necessity to add the envelope data creates an overhead that increases the data transmission rate that must be achieved to effect timely transmission of a signal, the successful implementation of a packet-switched voice service depends on achieving a balance among the effects of

- Handling overhead on transmission speed requirements
- Transmission delays

- Effects of packet loss on the quality of the extracted voice signals

As will be described, the tradeoffs among these performance characteristics are determined by the sizes of the packet header and payload.

Header The information that must be appended to each data segment to be transmitted over a packet-switched network must be sufficient to support unattended routing, handling, and node-to-node transmission of the packet across the network, as well as reconstruction of the original data set from its segments at the destination. Consequently, packet headers may comprise a large number of bits relative to the minimum data segment size generated by the voice codec. For example, as shown in Table 2-2 the header for an IP packet comprises a total of 160 bits, or 20 bytes. User datagram protocol (UDP) and real-time transport protocol (RTP) controls create the need for an additional 8 and 12 bytes of header information, respectively, bringing the total number of bits that are needed for packet headers to 320.

Insertion of the minimum-sized segments produced by voice codecs like those shown earlier in Table 2-1 into such relatively large envelopes

TABLE 2-2

Information Fields and Sizes in an IP Header

Information Element	Number of Bits
Version	4
IP header length	4
Type of service	8
Total packet length	8
Datagram identifier	24
Flags (3)	3
Segment offset	13
Time to live	8
Protocol identification	8
Header check-sum	16
Source IP address	32
Destination IP address	32
Total	160

can, then, result in prohibitively large handling overheads. For example, use of a 320-bit envelope for an 80-bit data segment created by a G.729 codec would increase the required data rate from 8000 to 40,000 bit/s, vitiating much of the bandwidth savings of the G.729 codec over the G.711 codec. To mitigate this deleterious effect on throughput efficiency, there are defined conventions for *header compression* that may be applied to reduce the size of the header. The RTP header compression convention, for example, reduces the IP, UDP, or RTP header requirement from 40 to either 2 or 4 bytes.

Payload Size The other means of reducing the handling overhead associated with packet-switched transport of voice data segments is to increase the size of the payload for each packet to include more than one of the minimum segments. This reduces the handling overhead and improves transmission efficiency.

As illustrated in Table 2-3, the data rate requirements for transmission of voice samples generated by different codecs vary substantially with the choice of payload size and application of header compression. Any efficiencies in transmission from increasing the payload sizes are, however, realized at the cost of an increase in packet latency and a greater effect of dropped packets on voice quality.

Increase in Packet Latency If, for example, the IP payload size for G.723.1 data is increased from the minimum of 189 bits shown in Table 2-1 to 378 bits, so that two encoded voice segments are transmitted in each packet, then the handling overhead is reduced from more than 169 to 85 percent and the required data rate is reduced from 16,947 to 11,655 bit/s. To do this, however, it will become necessary to wait for two 30-ms voice segments to be encoded before forwarding the packet, thereby increasing packet latency to 67.5 from 37.5 ms.

Greater Effect of Dropped Packets on Voice Quality Continuing the G.723.1 codec example, when only one segment per packet is transmitted, a dropped packet results in a gap of 30 ms in the sampled voice data, which represents about the duration of an articulated phoneme. In this case, a dropped packet compensated by packet loss concealment will result in noticeable speech distortion but will not affect intelligibility. When there are two segments per packet, each dropped packet results in a gap of 60 ms in the sampled voice, representing the duration of some syllables. In this case, packet loss concealment will be ineffective, and the dropped packets will begin to degrade voice intelligibility. More gen-

TABLE 2-3

Effects of Payload Sizing and Header Compression on Data Rate Requirements

Configuration		No Header Compression		2-byte RTP Header		5-byte RTP, UDP, IP Header	
Codec	**Payload**	**Data Rate, kbit/s**	**Handling Overhead**	**Data Rate, kbit/s**	**Handling Overhead**	**Data Rate, kbit/s**	**Handling Overhead**
G.729	One 10-ms sample	40	4.0	32	3.0	12.0	0.5
	Two 10-ms samples	24	2.0	20	1.5	10.0	0.25
	Four 10-ms samples	16	1.0	14	0.75	9.0	0.125
G.711	One 10-ms sample	96	0.5	88	0.375	68.0	0.0625
	Two 10-ms samples	80	0.25	76	0.1875	66.0	0.03125
G.723.1	One 30-ms sample*	16	2.025	8.05	1.519	5.45	0.03165
	One 30-ms sample†	17	1.693	8.00	1.27	6.46	0.02646

*5300 bit/s.

†6300 bit/s.

Note: Handling overhead (HO) is expressed as the ratio of the amount of handling data to the amount of injected data, so that the volume of injected data is inflated by the factor (1 + HO).

erally, because the effects of dropping *n* consecutive voice samples on speech distortion and intelligibility is always greater than dropping 1 voice sample, increasing payload sizes will exacerbate the effect of dropped packet rates on voice quality.

Jitter Buffer

In a voice service the digitized voice samples must be presented to the codec decoder in such a way that the next sample in a stream is present for processing by the time the decoder is finished with its immediate predecessor. Such a requirement severely constrains the amount of jitter that can be tolerated in a packet-switched service without having to gap the samples. When jitter results in an interarrival time between the packets carrying consecutive samples that is greater than the time required to re-create the waveform from a sample, the decoder has no option but to continue to function without the next sample information.

The effect of jitter on dropped packet rates implies that the incidence of dropped packets measured for a packet-switched voice service will be greater than that measured for the underlying packet-switched transport. It also eliminates jitter from the list of essential descriptors of intrinsic quality of a packet-switched voice service, because the effects of jitter will be manifested as an increase in dropped frame rates.

To make the generation of continuous analog voice signal at the distant end less susceptible to variations in the time of arrival of packets across the network, codecs used for packet-switched voice services have provisions for queuing a number of segments of digitized voice before decoding starts. This has the effect of increasing the magnitude of the interarrival time between samples that can be tolerated without gaping the voice samples by the amount of time it takes the decoder to clear the queue. (See App. B for details.)

The buffer that holds the queued segments is called a *jitter buffer.* The employment of such jitter buffers effectively defines the relationship between jitter in a digitized voice stream and dropped frame rates, trading off dropped frame probabilities against increases in transmission delays defined by the size of the jitter buffer. The *amount of difference in delay that can be tolerated* therefore becomes the essential descriptor of intrinsic quality that supplants jitter in the case of a packet-switched voice service.

Implications

Taken together the preceding discussions demonstrate that specification of the transport protocol and voice codec (e.g., VoIP using a G.729 codec) is by itself an inadequate characterization of a packet-switched voice service for purposes of measuring and reporting characteristics that are likely to affect user perception of the quality of that service. Rather, any system description must as a minimum also specify:

1. Whether silence suppression is activated and the characteristics of comfort noise used in conjunction with silence suppression
2. The length of the samples of voice digitized by the codec and the number of samples in the payload for each packet
3. Whether header compression is used
4. The size of the jitter buffer used by the codec decoder

CHAPTER 3

Quality Expectations and Requirements

Service Models

As suggested by the discussions in Chaps. 1 and 2, there are many possibilities for creating packet-switched voice services, defined by the packet-switched network in which the service is implemented, the type of codec employed, and the configuration options selected for that codec. There is, in addition, the question of the envisioned use of the voice service, which will shape the expectations of the users of the service. For purposes of illustration, we will consider three variants of service usage that will cover the spectrum of possibilities and serve to highlight the possible differences in user expectations conditioned by those services. The three service models are hybrid transport, packet-switched telephony, and interactive multimedia exchange.

Hybrid Transport (tpt)

In this model, the packet-switched network is used for transport of long-distance telephone calls completed across the public switched telephone network (PSTN). Call attempts are circuit-switched until the voice signals are injected into a *gateway* to a packet-switched network through which they are transported to a distant gateway. At the distant gateway they are then extracted for onward delivery to the destination station via circuit-switched terminations. Under this service model, the long-distance transport networks for voice and data are merged into a single-mode packet-switched network, such as the Internet, thereby achieving economies of scale in operation and maintenance, and possibly some reduction in costs of long-distance transport capacity.

In the case of a hybrid transport service, nothing is different about the way users originate and answer calls, and there is no apparent benefit to the users for any resultant change in quality of their long-distance telephone services. Employment of hybrid transport must, therefore, be transparent to users, supporting perceived quality of service that is not noticeably different from that achieved in the comparable circuit-switched services.

In practical terms, this means that in a hybrid transport voice service:

1. The expected user perception of voice quality must be as good or better than that for the circuit-switched service or, at worst, less by an amount that is not *operationally significant.*

2. There must be a very infrequent occurrence of impairments, impediments to natural conversation, or line conditions that are rarely manifested in the circuit-switched service.
3. There must be no operationally significant increase in the perceived incidence or severity for any of the familiar impairments.
4. All uses of today's circuit-switched voice services, including, for example, transmission of fax and dial-up data and end-to-end connections with wireless mobile services, must be supported.

Packet-Switched Telephony (tel)

In this model, the packet-switched telephone service is hosted on an existing private or virtual private packet-switched data network. The gateways into the packet-switched network are customer-owned and already in place at nodes of the private data communications network. The voice service is overlaid onto this network, either by use of voice gateways interfacing directly with the customer private branch exchanges (PBXs), or by use of direct session initiation protocol (SIP) terminations to voice stations implementing the selected codec.

Because packet-switched telephony, as we have defined it here, will be implemented on private or virtual private networks, users of the service will be more likely to know that it is somehow different from their familiar circuit-switched service, particularly when it is terminated via SIP telephone sets. Experience with similar replacements of circuit-switched services with satellite-based services suggests that users will in this case be somewhat more tolerant of noticeable differences between the circuit-switched and packet-switched services, as long as two conditions are satisfied. The first condition is that it must be widely known that substantial cost savings or other tangible benefits to the company are being realized by using packet-switched telephony. Otherwise, the users will expect the new service to be as good as or better than the old and will tend to perceive any differences as degradations in quality, even though those differences might otherwise not be expected to have a substantial effect on voice quality or connection usability. The second condition is that the differences between packet-switched and circuit-switched telephony do not substantially increase the incidence of calls that are rated by users as "unusable," "difficult," or "irritating."

In addition, the users of packet-switched telephony will expect accommodation of the other uses of circuit-switched voice services, such as fax and dial-up data. The ideal arrangement for this in the envisioned environment for packet-switched telephony would be inclusion of embedded handlers, which are capable of demodulating fax and acoustic data signals at the origin, transmitting the content as data packets, and remodulating the data at the destination. However, the users would probably be content were such accommodation to simply require installation of the devices on analog lines, just as is done today in locations served by digital voice telephony behind the PBX.

In practical terms, this implies that for a packet-switched telephone service:

1. The expected user perception of quality must be no worse than that for the worst comparable circuit-switched service for which users have reported the service quality as being satisfactory.
2. The expected proportion of calls that will be rated "unusable," "difficult," or "irritating" must not exceed known tolerable limits for the comparable circuit-switched service.
3. There must be accommodation of transmission and reception of fax and dial-up data in the environment served. Such accommodation does not, however, have to be implemented in the packet-switched telephone service.

Interactive Multimedia Exchange (ime)

In this model, the voice service complements and enhances other exchanges of information via the packet-switched data network. Interactive multimedia exchange via the Internet would allow users, for example, opportunities to engage in interactive exchanges of voice while browsing a web-hosted catalog to elicit more detailed information about a particular item whose image and descriptive textual material are simultaneously displayed on the user's computer screen. In this kind of packet-switched voice service, the voice codecs are hosted on the computers that are supporting exchanges of text files and image data.

As an overlay on an existing packet-switched data network, interactive multimedia exchange will support creation of attractive new capabilities in the host medium, such as the web-shopping feature just described, where the user is assisted by live dialogues with salespersons who would answer questions as the user browses a web-hosted catalog. Others include IP-hosted videoconferencing, picture telephones implemented on personal computers (PCs), and use of a PC as the station set for general telephony.

The principal benefits realized by users of interactive multimedia exchange will, therefore, be access to telecommunications capabilities that either do not currently exist, or do exist but are only crudely and ineffectively implemented.

Experience shows that when a service supports new capabilities for which there are no existing comparable capabilities, users tend to be much less demanding, accepting in the new service quality that which would be deemed to be unacceptably poor in other applications. For example, users of cellular telephone services accept connectivity and quality that would be completely unsatisfactory in their home service as an unfortunate, but inescapable, inconvenience. Similarly, the precursor to interactive multimedia exchange, IP telephone service implemented on PC microphones and speakers, has been placed in use, without complaint, by persons who are happy to suffer the very low quality for the opportunity to use the Internet to avoid the high cost of International telephone calls.

This implies that user expectations and requirements for interactive multimedia exchange services will be altogether different from their expectations for the other kinds of packet-switched voice services. Rather than expecting quality that compares favorably with circuit-switched voice services, users will be concerned with the adequacy of the voice service in each application. In particular, this means that in an interactive multimedia voice exchange service the following are necessary:

1. The voice heard must be clear and undistorted enough to be *intelligible* to a listener who is not straining to hear.
2. Transmission of voice must preserve natural speech rhythms, inflections, and cadences.
3. Round-trip conversational delay, comprising the time lapsed between articulation of a thought and hearing the distant speaker response to that thought, must be stable and not great enough to cause irritation or disruption of the flow of ideas.

In addition, because the voice service is in this case overlaid on a packet-switched network already handling data exchanges, there is no necessity to accommodate transmission of fax or acoustic data via an interactive multimedia voice exchange service.

Summary

The preceding characterizations of the likely user expectations for the three service models examined are summarized in Table 3-1.

TABLE 3-1

Likely User Expectations as a Function of Type of Packet-Switched Voice Service

Service Model	Voice Quality	UDI	Unusual Impairments	Fax/Acoustic Data
Hybrid transport	Not operationally significantly worse than PSTN		Infrequent relative to PSTN	Transparent handling
Packet-switched telephony	Satisfactory by PSTN standards		Neither frequent nor quality impacting enough to render voice quality or UDI unacceptable	Accommodation within the environment
Interactive multimedia exchange	Intelligible by listening tests	PSTN standards		No requirement

Implications for Codec Selection

The best codec for a particular service model will ultimately depend on the relationships between intrinsic and perceived measures of quality determined by the characteristics of codecs and their configurations as described earlier, in Chap. 2. However, analysis of sources of impairments in packet-switched services just described in light of known characteristics of codecs and the requirements set forth in Table 3-1 does support some conclusions that reduce the number of possibilities that need to be considered for each service model. These are as follows:

1. *The only codec that is viable for the hybrid transport service in the near term is G.711, running without VAD/VAC.* As has been demonstrated both by analysis and by actual testing, the CELP coding used in the G.723.1 and G.729 codecs preserves phase but produces too much amplitude jitter to support the phase- and amplitude-modulation schemes used in high-speed acoustic data modems. This means that the fastest fax and dial-up data transmission speeds that can be reliably achieved using CELP is 7200 bit/s, as compared to the 14,400 bit/s routinely achieved with fax cards and speeds up to 56,000 bit/s achieved with V.90 data modems. Since today's PSTN provides no capability for separate handling of fax and acoustic data calls, a hybrid service based on the G.729 or G.723.1 codec transport across the packet-switched network cannot meet the requirement described in Table 3-1 in the near term. The restriction to the use of the G.711 codec without VAD/VAC is mandated by the same limitation. When there are no provisions for differential handling of fax and data modem signals, application of voice activity detection will result in premature disconnects of acoustic modem transmissions, because a long, standing signal without drops in power will eventually be classified as high background noise and suppressed. Moreover, even were there no problem with transmission of acoustic data signals, the use of VAC with the G.711 codec would have to be carefully qualified before it could be considered for use in a hybrid transport service. Such qualification would require, in particular, demonstration that the exposure to noise on speech would not be great enough to create an appreciable probability that PSTN users would notice it as a new impairment in their telephone service.

2. *Among the three codecs that we have examined, the best candidate for use in an Internet telephony service is G.729.* Subjective tests comparing user perception of quality of voice transmissions via the PSTN with VoIP using the G.729 codec have consistently shown that, absent effects of dropped frames and excessive round-trip delay, user perception of G.729 service is not operationally different from that of the PSTN service. Since the PSTN uses the G.711 codec without VAD/VAC for digital transport, this means that G.729 is a viable substitute for G.711 in any context in which the handling of acoustic data signals is not an issue. In such an environment G.729 will also be a more cost-effective alternative, because the G.711 service would have to use VAD/VAC to achieve any bandwidth reductions, whereas the G.729 could be used without VAD/VAC to achieve significant reductions in capacity requirements. The G.729 codec will therefore offer voice service whose quality is as good as or better than G.711 at a lesser cost for capacity. In the opposite direction, even if it is assumed that the G.723.1 codec supports voice quality that is otherwise not operationally different from a similar G.729 service, the tradeoff in this case in a modest reduction in the capacity requirements against a substantial increase in exposure to unacceptable round-trip delays.

3. *The G.723.1 codec is a viable candidate for use in a voice service only in an environment in which it is critical to reduce bandwidth requirements* and *there in inherent resiliency to delays between oral stimuli and aural responses.* As shown in Table 2-1, the G.723.1 codec realizes a reduction in base signal data rates to 5300 or 6300 bit/s from the 8000 bit/s required for the G.729 codec, but at a cost of using 30-ms samples instead of 10-ms samples. With the smallest jitter buffer, this difference will add more than 110 ms to the round-trip delay, while substantially reducing resiliency to jitter and dropped packet rates over a G.729 codec carrying 10-ms voice samples in each packet. Against this exposure to degradation of perceived QoS, the G.723.1 codec in a VoIP application will achieve a reduction in capacity requirements over a similar G.729 service of about 60 percent when full headers are used and 34 percent when header compression is used.

CHAPTER 4

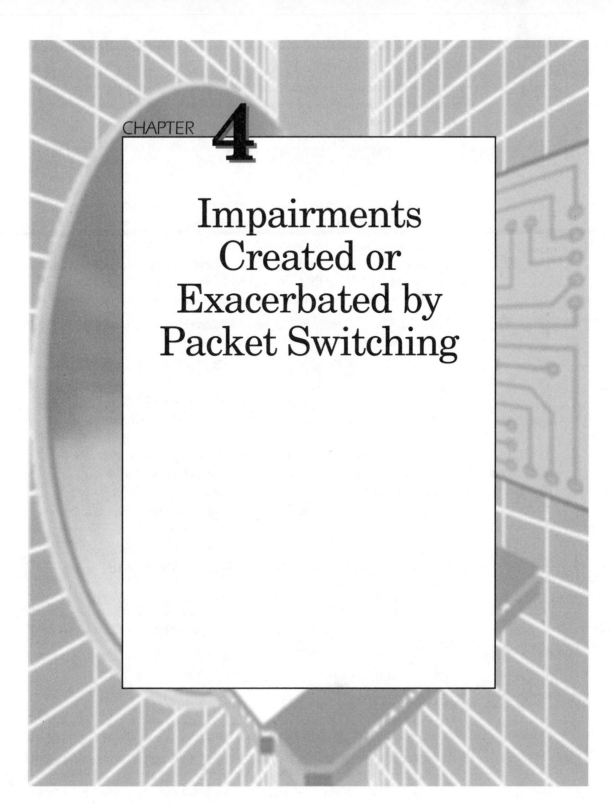

Impairments Created or Exacerbated by Packet Switching

As suggested by the description in Chap. 3 of likely user expectations of the various packet-switched voice services examined, the users' perception of quality of those services will ultimately depend on:

1. Differences in incidence or severity of impairments experienced by users between the packet-switched service and more familiar, circuit-switched services
2. Incidence of problems or impairments in the packet-switched service that are rarely manifested in circuit-switched services

Since speech power levels and common line noise will not be affected by digitization, the impairments or problems manifested to users of packet-switched voice services whose incidence or severity might exhibit a noticeable difference from those in familiar circuit-switched services include:

- *Noise on speech.* A phenomenon in which there is a marked contrast between the background noise heard when the distant party is speaking and when the channel is quiet, produced by silence suppression.
- *Echo.* The reflection of a speaker's speech signal back to the origin with enough power and delay to make it audible and perceptible as speech.
- *Speech distortion.* Deformations of natural speech waveforms that produce sounds that cannot be articulated by human speakers.
- *Voice clipping.* Loss of beginning or ending sounds of words at the distant end.
- *Disruption of conversational rhythms.* Caused by inordinately long pauses between the time a user stops talking and the time that user hears a response.

The characteristics of a packet-switched voice service that may affect the incidence or severity of each of these impairments are described briefly below.

Noise on Speech

The impairment that users describe generally as "noise on the line" is largely generated in the segments of end-to-end voice connections that utilize analog transmission. Thus, in hybrid transport service or packet-switched telephone service, neither the incidence nor the severity of noise will be greater than in the familiar circuit-switched services to which users will compare them. Moreover, some of the codecs, such as G.729 and G.721.1, are based on methods whereby received segments of voice signals are processed to determine the best fit with ideal segments, and the name of the ideal segment is transmitted for perfect regeneration by the decoder. Such a process effectively filters out low-level noise that has been added to the voice waveform and suppresses noise altogether in quiet segments, resulting in, if anything, a reduction in the user's perception of noise on the line.

Instead of noise, itself, then, the noise-associated problem in any digitized voice service is a phenomenon, called here *noise on speech,* in which there is a noticeable contrast between the noise heard when no one is talking and that heard as background noise when the distant speaker is talking. This phenomenon is invariably associated with VAC and manifested in either of two ways:

1. When VAC is used without insertion of comfort noise by the decoder, the presence of even subliminal line noise in speech signals may be enough to make it appear that the call has been disconnected when the distant speaker stops talking. When manifested in this way, the expected effect of noise on speech is some increase in the incidence of disruptions of conversational rhythms.
2. When VAC is used with comfort noise to avoid the "deep null" phenomenon (see Chap. 2), a very loud background noise when there is speech on the line may be great enough to create the perception of noise that occurs only when the distant person is speaking. When manifested in this way, the dominant reaction of the user will be to gauge the louder noise perceived against the speech. Therefore, noise on speech as a phenomenon, if anything, reduces the user's perception of the severity of noise on the connection.

Since there is no benefit from using VAC on circuit-established connections, noise on speech is one of those impairments that is rarely encountered in familiar circuit-switched voice services. Thus, even though it has little direct effect on the quality of calls, noise on speech can affect user satisfaction with a service.

Echo

In a telephone network, echo is generated at the termination side of connections by reflection of the transmitted voice energy across the hybrids that convert four-wire transport links into two-wire subscriber loops. The likelihood that such reflected energy will be perceptible to the distant talker depends both on how much the echoed signal is attenuated by the time it returns to the speaker (echo path loss) and the time between the original utterance and its echo (echo path delay). In the service models that we have examined, the only places that echo might be generated are in the access and termination segments of a hybrid transport service or the access and termination of a packet-switched telephone service through PBXs. In the cases where there is exposure to echo, however, the handling and transmission times in a packet-switched voice service may substantially increase the echo path delay over that expected in a comparable circuit-switched service. Such exposure is of a particular concern, because the increase in echo path delay will have two effects:

1. It will exacerbate the severity of echo that was experienced over the circuit-switched routes.
2. It will increase the expected incidence of perceptible echo by rendering audible any echo that was there, but not delayed enough to be perceptible on circuit-switched routes.

Characterization and control of echo thus becomes a critical issue in designing and implementing hybrid transport and packet-switched telephone services.

Speech Distortion

In attempting to gauge the likely user perception of quality of a voice service, there is a temptation to equate speech distortion with fidelity of the speech signals produced through codecs. However, *speech distortion* in the sense used here is something that goes well beyond the subtle differences in tonal quality and audible frequency ranges gauged in tests of codecs under laboratory conditions, without any of the interfaces in real-world, end-to-end connections. Rather, as manifested in the day-to-day use of telephone service, speech distortion is a deformation of the distant speaker's waveform that has the effect of making that speaker's voice sound unnatural, because it contains sounds that would never be heard in face-to-face conversation. Such distortions are variously described as making the voice sound "raspy," "whispy," "feathery," "warbly," "like someone gargling," "muddy," etc.

These kinds of distortions have their origins in the digitization or digital transport of the voice signals. Possible causes in any network include, for example:

1. *PCM quantizing noise that is not adequately filtered by signal attenuation or masked by ambient line noise.* When a PCM codec like G.711 encodes a pure voice signal at a high level without ambient noise, the differences between the encoded amplitude values that are transmitted and the actual amplitude values sampled produce what is called *quantizing noise.* If that signal is regenerated without attenuation or addition of white noise to the resultant analog signal, the reconstructed voice sounds like someone with a strange hoarseness.

2. *A high bit error rate incurred in the transmission of a digital voice signal.* In any digitized voice signal, a bit error incurred in transmission will result in deformation of the speech waveform generated by the decoder from what was encoded. Depending on where and how often they occur, such deformations can range from having little effect on what the user hears to producing something that is intelligible, but unlike anything that any human being is capable of uttering.

3. *Inordinately high levels in the analog signals fed into a codec.* One of the most common causes of what is perceived to be speech distortion in today's circuit-switched voice services utilizing PCM codecs occurs when the peak amplitude of the analog signal fed into the codec exceeds the maximum value that it was designed to handle. When this happens, the higher and lower amplitude values are assigned the maximum and minimum values in the codec, producing representations of the analog waveform in which the extreme portions have been squared off. When these squared-off waveforms are fed into the decoder at the distant end, the results can be truly bizarre. For example, a DTMF-# signal becomes a clearly recognizable DTMF-#/DTMF-3 combination, and voice becomes very brittle, sounding like s's and t's are being whistled, rather than spoken. (DTMF stands for dual-tone multiple frequency.)

In addition to the examples of speech distortion just described, which may be experienced in any telephone service using digital transport, dropped frames in a packet-switched voice service will produce waveform deformations that result in perceptible speech distortion. The incidence and severity of speech distortion attributable to dropped frames in a packet-switched voice service will vary with

1. Size of smallest sample of voice encoded in the codec
2. Number of voice samples per packet

3. Dropped packet rate
4. Use and efficacy of packet loss concealment

The size of the smallest sample of voice encoded in the codec, together with the number of samples per packet will determine how much of the original voice signal is lost when a packet is dropped, thereby defining the potential severity of waveform deformation due to missing packets. The particular method of packet loss concealment, together with the dropped packet rate will then determine the likely perception of the severity of the resultant speech distortion.

Another characteristic of the codec used that will affect the incidence and severity of speech distortion in any service is the type of encoding used (PCM, ADPCM, CELP, etc.), which will determine how much waveform distortion will result from the common causes described earlier. Notice, however, that what is absent from the list of codec characteristics enumerated here are (1) the codec data rate and (2) any of the measures of waveform distortion used for benchmarking codec quality. The reason is that neither of these seems to have any utility in predicting user perception of speech distortion. Tests that we have run in the operational environment consistently show, for example, that voice transmitted using a G.729 codec, running at only 8000 bit/s with no dropped frames, is rated as good or better than voice signals transmitted using a G.711 codec running at 64,000 bit/s. In similar tests in which the 23-tone test for waveform distortion produced signal to distortion ratios in the expected range of 35 to 39 decibels (dB) for a 64,000 PCM signal (G.711 codec), the values for the G.729 codec were less than 20 dB and sometimes as low as 2.5 dB for calls rated by users as having no speech distortion.

Results like these suggest that we should use a note of caution when trying to assess user perception of quality of digitized voice services. Although greater data transmission rates generally result in greater signal fidelity, the human ear and brain process waveforms in a way that is resilient to wide ranges of what might otherwise be measured as extreme deformations. As a consequence, the results of subjective tests of codecs in a laboratory environment, absent any of the interfaces and effects in end-to-end transmission, tend to exaggerate differences among different codecs with respect to expected user perception of quality. In the laboratory environment, careful listeners will often base their evaluations on subtle differences in tonal quality that are imperceptible when gauged against the kinds of speech distortion to which users are exposed in the real world. Or, to put it another way, 25 dBrnC of white noise can totally mask many of the differences that were detected and scored when codecs were benchmarked in the laboratory. (dBrnC is a measure of noise power whereby 10 is very quiet and 40 would begin to overpower speech. dB = decibels; rn = relative to −90 dBm; C = C-message weighting applied in the noise energy calculation.) This means that any comparative test data for voice codecs must be carefully qualified to assure that the tests were conducted in a way that at least simulated a real-world operating environment.

Voice Clipping

Voice clipping, in the sense used here, occurs only when VAC is employed. It results from a slow response or maladjustment of the VAD (voice activity detection) processors which prevents transmission of the beginning of words, because the VAD is slow to recognize them, or the soft endings of words, because the VAC prematurely cuts off transmission. Unlike undisguised dropped frames, which produce randomly distributed gaps in the reconstructed speech signals, the voice clipping that occurs when VAC processors do not react quickly enough consistently occurs at the beginning and ending of words, leaving some spoken words undistorted, but incomplete. Because of the redundancy in spoken language, this kind of word clipping does not greatly affect intelligibility of speech. If it occurs consistently in a telephone conversation, however, it can become one of those characteristics that renders an otherwise clear, unimpaired telephone call unacceptable.

Disruption of Conversational Rhythms

Because transmissions of packets across a packet-switched network do not necessarily take the same route, transmission delays are not necessarily symmetric. This means that it is possible for the delay in transmitting a packet from point A to point B to be substantially different from the delay in sending a packet from point B to point A. In the analysis of packet-switched data services, this feature is an important consideration, because it is the origin-to-destination delay that affects the performance of the data transport.

In a voice service, however, user perception of quality is affected by transmission delays in ways that depend only on the round-trip delay. As already described, such round-trip delays substantially affect user perception of the incidence and/or severity of echo.

Another deleterious effect is *conversational disruption,* which can result from inordinately long round-trip delays. In normal face-to-face conversations the exchanges of speech take on a rhythm that sets the conversationalists' expectations of the time that will lapse between the last word spoken in the articulation of something expected to produce a response and hearing the first word of the response. Over a telecommunications connection, the perceived conversational delay of this kind is the round-trip delay. Any substantial increase in round-trip delay from that experienced or expected on a circuit-switched service may, therefore, result in unexpectedly long delays between the articulated stimulus and the audible response. Such unexpected delays are well known to be disconcerting, resulting in what are perceived to be "awkward silences," or causing unwarranted attempts to reiterate the stimulus for a response. Since the likelihood of such disruptions to conversational rhythms increases with the magnitude of the round-trip delay, increases in transmission and/or handling time in a telecommunications system will increase the likelihood that users will find calls whose quality is otherwise excellent to be "unusable," "difficult," or "irritating."

PART 2

Measurement and Evaluation of Voice Quality

Chapter 2 shows that packet switching may increase telecommunications capacity requirements for voice services. Chapter 4 shows that packet switching will change the characteristics of voice transmissions and exchanges in ways that may substantially degrade users' perception of voice quality. Because of these possibilities, neither service providers nor their customers are going to be comfortable with the idea of moving on to packet-switched voice services until there is a clear determination of what it takes to make those services both cost-effective and acceptable to users.

A key to resolving this issue is development of a reliable means of predicting user perception of packet-switched voice quality from parameters that objectively describe connection quality. Such a predictor would, for example, accurately translate the effects of packet latency, jitter, and dropped frames into reliable indicators of likely subjective assessments of voice quality by users of packet-switched telephony using various different codecs. It would similarly translate measures of packet-switching effects together with familiar measures of circuit-switched voice connections to produce analogous predictions for a hybrid transport service. The ideal capability of this kind would be suitable for use in two applications:

1. In development efforts, to support determination of performance characteristics that must be achieved by packet-switched networks to assure acceptable voice quality for various different services and configurations of a particular service
2. Later, without change, to characterize expected voice quality for prospective users of developed packet-switched services

The purpose of this part of the book is to describe and compare alternate ways and means of producing such a predictor.

CHAPTER 5

Voice Quality

The principal goal in measuring and evaluating the quality of voice in packet-switched services is the development of predictors of user perception of voice quality that credibly and reliably reflect the specific effects of packet switching described in Chap. 4. This chapter lays the foundations for that quest by

- Describing the processes by which users assess telephonic voice quality
- Defining measures of voice quality that appropriately recognize those processes

User Assessment of Voice Quality

When users talk about the *quality* of voice carried over an interactive voice connection, they are generally trying to describe their reaction to, or satisfaction with, one of two attributes of a voice service:

- *Connection quality.* Determined by what is heard over the connection
- *Connection usability.* Determined by what is experienced in conversational exchanges over the connection

In listening to a connection to assess connection quality, users will, for example, be conscious of whether there is perceptible echo; the distant speakers are easily heard, readily understood, and sound "natural"; the connection is degraded by distracting impairments; and the voice fidelity is good enough to enable recognition of distant speakers and to detect nuances in their articulation.

In contrast, when assessing the usability of a telephonic voice connection, users will determine whether the overall effect of all impairments and conditions on that connection was severe enough to interfere with natural conversational rhythms and speech patterns or was otherwise distracting to the flow of information.

User perception of the quality of a voice service with respect to these two attributes is almost a primitive notion, constituting something that is inescapably subjective and dependent on the tastes, dislikes, and expectations of each individual user. However, as we will describe, the mechanisms by which each user will ultimately arrive at such a subjective evaluation of a voice quality are nearly universal.

Connection Quality

In the case of connection quality the speech signals carried over a voice connection will comprise two parts:

1. That which emulates natural, face-to-face conversations with sufficient fidelity to enable comprehensible exchanges of information
2. The differences between the emulated face-to-face conversation and what is heard

The differences are created in the processes by which natural speech is turned into electrical signals, encoded, transmitted, decoded, and received as acoustic waveforms at the distant end of a voice connection. They are perceived as what is "left over" or different once the human listener has extracted what is natural and expected from what is heard. They result in artifacts that may be perceived but are ignored in the cognitive process that enables us to understand spoken language.

Depending on their magnitude, those artifacts may be subliminally filtered or consciously recognized. Some of the consciously recognized artifacts such as noise and echo are simply ignored in developing an understanding of what is being said. Others, such as unnatural distortions of voice waveforms, partial clipping of words, and speech levels that are difficult to hear cannot be ignored, because the words affected must be mentally re-created to be understood.

Either way, however, those differences are *manifestations* of underlying conditions degrading the fidelity of voice connections that can be readily recognized, distinguished, and described by, or described for, users. In evaluating quality of voice connections, users will synthesize their experiences over a number of calls to form a subjective impression of the incidence and severity of those manifestations. Their assessment of the overall quality of connections made through the service will then reflect a comparison of that subjective impression with expectations that have been conditioned by years of use of voice services.

Connection Usability

The user evaluation of connection usability is the result of a similar process by which users form a subjective impression over a number of calls. What happens in this case, however, is that users synthesize their experiences with the use of a voice service from what they remember to be bad calls, on which the parties to the call

- Were forced to curtail the conversation, possibly reinitiating the call to try to get a better connection
- Maintained the connection, but had to repeat or ask for repetition of information, speak louder or slower than normal, or adopt unnatural conversational patterns in order to hold the conversation
- Found the conditions on the connection on the whole to be so distracting as to interfere with their ability to focus on what was being said

The frequency with which such calls are encountered is a major, independent determinant of users' assessments of the acceptability of a voice service. Regardless of how good the quality was on other calls, users will remember such bad calls. They will become dissatisfied with the service when their subjective estimates of the frequency with which bad calls occur suggest that the service offers too much exposure to the remembered conditions.

Measure of Connection Quality

A viable measure of connection quality is one that will provide some basis for addressing the following users' concern: "Will what I hear over the voice connections sound all right?"

Definition

There are a variety of measures of quality that have been proposed or employed for addressing this question over the years, including such anachronisms as scores on diagnostic rhyme tests, which gauge intelligibility of voice signals. Nearly all such measures today, however, are based on responses elicited from test subjects by

1. Defining categories of possible subjective descriptions of the quality of a voice connection, such as "excellent," "good," "fair," "poor"
2. Conducting tests in which participants listen to voice signals and assign subjective responses to samples

In early testing of this kind, the measures used were simply proportions of the sample responses that fell, or were expected to fall, into certain categories. Thus, for example, outputs from the earliest attempts to predict voice quality with the so-called loss/noise grade of service model developed by AT&T in the mid 1960s were simply expected percentages:

%GoB. Represents the percentage of calls rated as "good" or better
%PoW. Represents the percentage of calls rated "poor" or worse

Mean Opinion Scores

The most widely used and accepted measure of user perception of voice today is, however, a *mean opinion score* (*MOS,* pronounced "moss"). Such a MOS is a value created by

- Assigning numerical values to each of the subjective descriptors
- Calculating the average of the corresponding numerical values

The particular values assigned to each category might, in fact, be anything. However, the convention adopted by the ITU in Recommendation P.800 (Ref. 3) is the five-value scale shown in Table 5-1. This table also shows other scales that have been used. The 0 to 4 scale in the table, for example, is the original quality scoring scheme used by AT&T in the 1960s. The fractional score scales are ones that have been allowed or explicitly adopted in various subjective tests.

Whatever scale is used, however, the presumption when describing the MOS is that the value represents what resulted, or would be expected to result, when

1. A large body of users is asked to report their subjective opinions of the quality of a number of voice connections on the scale specified
2. The numerical values assigned to all those reports is averaged

TABLE 5-1

Opinion Scales for Subjective Tests

	AT&T 1960s	ITU P.800 1996	Fractional Scores		Interpretation
Excellent	4	5	3.5	4.5	Very good
Good	3	4	2.5	3.5	Fair to good
Fair	2	3	1.5	1.5	Fair to poor
Poor	1	2			
Unsatisfactory/bad	0	1			

Pitfalls in Interpretation of MOS

Because of the formal definition and widespread use of the abbreviation MOS, there is a tendency to assume that a mean opinion score is a stable, immutable measure of voice quality which always means the same thing, just as a noise measurement of 20 dBrnC always means the same thing. However, in the case of mean opinion scores, a particular value, such as 3.75, cannot be meaningfully interpreted without at least a description of the context of its calculation, and probably not even then without a known comparative reference for that context.

We would like to think that the mean opinion score reflects the likely user perception of voice quality. In one sense, this is true, because a MOS value is presumed to represent something that is created by the process described earlier, representing a body of subjective assessments of voice quality by a group of test subjects. Thus, a greater incidence of favorable opinions should result in a higher mean opinion score, and it may be safe to infer that a higher mean opinion score for a particular service implies that there is a greater likelihood that the general population of users will find the voice quality satisfactory.

However, the problem is that any particular value of MOS by itself is largely meaningless. Suppose, for example, that we report that a subjective test was run and the mean opinion score was 3.75. One thing we need to know before we can even begin to interpret that score is what the scale is. For this we need to know both precisely what quality descriptors were defined and what numerical values were assigned. The question of the numerical values is critical, because, as shown in Table 5-1 there has been a historical shift from an earlier convention in telephony of assigning "excellent" a value of 4, with integer steps for the other quality descriptors, to a later ITU standard in which the best value assigned is 5. Unless we know the scale, then, our 3.75 may reflect an overall quality describable as "very good" or quality that is less than "good."

There are, moreover, mean opinion scores that are quoted for which the underlying subjective descriptors are not qualitative ones, like "excellent" and "good," but are other choices, such as "no effort required to hear" or "no appreciable effort required." The results from tests eliciting these kinds of responses are reported MOSs, but clearly must be interpreted differently from a MOS based on quality descriptors.

Moreover, these are but the obvious requirements for specifications of how the tests were run before we can even begin to ascribe meaning to MOS values. Other subtle differences in test protocols that must be discriminated because they have a strong effect on the value of the mean opinion score that will come out of the test include the convention for handling shades of evaluation, the method of sampling, and the type of test.

Convention for Handling Shades of Evaluation In the assignment of scores, there are at least two possible conventions for handling responses indicating an uncertainty in assessment. A test participant being interviewed might, for example, try to report that the quality of a particular call was "good-to-fair." Another, writing in responses, might be hard pressed to try to explain that the quality of the call just evaluated was not as good as the last call reported as being "excellent," but was noticeably better than the last call reported as "good." In a particular test, the expected tension created by the test participants' compunction to grade more finely than allowed can be ignored or accommodated in the test protocol. If it is ignored, the scoring is characterized as being *forced*, so that one response is mandated. The alternative is to accommodate the tension by allowing for compound subjective responses and/or fractional opinion scores like those shown in Table 5-1. In this case, responses like "excellent to good" or "very good" become admissible and may be assigned fractional opinion scores like 3.5 or 4.5, depending on the scale. Results from subjective tests that were forced on one side of a call, but allowed for fractional scoring on the other, have demonstrated that the MOS estimated from forced data may vary by as much as 0.30 from that obtained for the identical service when fractional reporting is allowed.

Method of Sampling In a subjective test the sampling of call quality of a particular service may be done by eliciting a single, or small number of, assessments from a large body of test subjects, or by using fewer test subjects who place repeated calls. Experience shows that in the case of repeated evaluations, the test subjects ultimately insist on not being forced and tend to report a much higher incidence of fractional scores. The result is that the MOS for a particular service will vary with the method of sampling used by as much as 0.25 to 0.30.

Type of Test In the arena of test and evaluation of voice quality there are essentially two different types of tests that are recognized as viable media for eliciting subjective evaluations and calculating MOS values. The first is a *listening test* in which test participants listen to standard voice recordings under conditions that are carefully controlled with respect to the volume at which the recordings are played, ambient room noise, and electrical conditions of telephonic devices and connections used. The second type is a *conversational test* in which the test participants engage in conversational exchanges conducted over the telephone. Conversational tests may be similarly conducted under controlled conditions in a laboratory environment or under actual operational conditions using operational networks, to expose the test participants to what service users will experience. Unless we know what type of test was conducted to produce the resulting mean opinion scores, we may as well ignore the results, because even comparative MOS values will be vastly different, for different types of tests. In the controlled listening tests, for example, the test participants will detect and appropriately grade down smaller differences in voice tone or quality that will be totally imperceptible to conversational users. Conversational test participants will be exposed to impairments and difficulties, such as echo and excessive delay, that are never manifested in a "listen-only" environment, but may entirely change a user's perception of overall quality of a connection independently of any assessment of how the call sounds. And, conversational tests in a laboratory environment may be based on simulations of impairments that do not accurately replicate their manifestations in operational networks.

So far, then, it should be clear that if we are going to be able to interpret the MOS derived from a particular test, we must, as a minimum, also discriminate the scale, type of test, method of sampling, and convention for handling shaded evaluations used in that test. However, even with all that information, we will not have an adequate basis for gleaning *information* as to likely user satisfaction from the MOS reported, unless the MOS for a particular service being evaluated is complemented with a similarly derived value of MOS for a known service. The reason is that absolute MOS values, even when they are derived from tests conducted under a particular, known test protocol, *are meaningless*. To say, for example, that the MOS for a particular service as evaluated in a controlled environment in a listening test with forced responses using a scale in which 5 denoted "excellent" was 4.3 reveals absolutely nothing. The value of 4.3 suggests that the service was reported to be better than "good," but not clearly "excellent," so it was probably all right, maybe. However, unless it is known how that value compares with similar evaluations of a familiar service that users find satisfactory, we will have no idea whatsoever as to whether the MOS of 4.3 will afford service that will meet customers' expectations. On the one hand, if that 4.3 value was reported by a group of test subjects known to have rated U.S. domestic long-distance service as a 4.1 in an identically structured test, then 4.3 is a good score, indeed. On the other hand, if the 4.3 value was reported by a group of test subjects whose rating of U.S. domestic service was 4.75, then that MOS value was a poor one. Without a reference value, then, the same MOS value may suggest that users will find the service better than one with which they are satisfied or the opposite conclusion, with no way of resolving the ambiguity.

Moreover, it should be cautioned that even with an adequate reference point, the comparisons may not support an accurate prediction of likely user satisfaction with voice quality, unless the MOS values can be confirmed to have come from commensurate distributions of opinion scores. For example, suppose that the 4.1 value for the U.S. domestic service resulted from a distribution of opinion scores in which 25 percent were 5, 60 percent were 4, and 15 percent were 3, but the 4.3 value resulted from a set of reports in which the users assessed 87 percent of the calls to be "excellent" and 13 percent to be "bad." Then the favorable comparison of the mean opinion scores would be totally misleading, because it would suggest that the evaluated service would be found to be better, when, in fact, that service would almost surely be rejected as totally unsatisfactory due to an unacceptably high incidence of "bad" calls.

Measurement of Connection Usability

In contrast to the question of how good transmitted voice sounds, which is addressed by measures of connection quality, the principal user concern with connection usability is expressed in the question, Will we be able to maintain natural, clearly understood conversational exchanges, without distraction, or interruptions caused by needs to adjust to the connection?

Definition

Unlike mean opinion scores, which are widely used to discuss connection quality, there are no widely used or accepted measures that have been defined to directly address this particular question. ITU P.800 does define a scheme for scoring listening effort with the descriptors shown in Table 5-2. This represents recognition, at least, of the important distinction between what is *heard* and what is *experienced* and the need for different kinds of measures to capture that distinction.

TABLE 5-2
Listening Effort Scale Defined in ITU P.800

Effort Required to Understand Meanings of Sentences	Score
Complete relaxation; no effort required	5
Attention necessary; no appreciable effort required	4
Moderate effort required	3
Considerable effort required	2
No meaning understood with any feasible effort	1

However, if the earlier description of the processes by which users assess connection usability is accurate, then what is needed in this context is something that defines bad calls in terms of overall experience and measures the frequency with which users are likely to perceive their occurrence. This kind of measure is exemplified by the proportion of calls that have been rated for overall effect of whatever was experienced regarding the ability to maintain a smooth, natural flow of conversation.

One scale for describing such effects is defined by the criteria shown in Table 5-3. These definitions are taken from the protocols for service attribute tests (SATs), which are described in Chap. 6. The associated measure, *P[UDI],* is defined as the proportion of calls that SAT subjects rated, or would have rated, as being "unusable," "difficult," or "irritating" according to that scale.

Although P[UDI] is specific to the service attribute test, it will be treated here as the exemplar for any analogous measures that can be used to gauge likely user perception of usability of voice connections. Such measures are distinguished by two characteristics:

1. Rather than reflecting what was heard over a voice connection that might *cause* disruptions to the flow of conversation, the subjective assessments reflect the *effects* of everything experienced on a connection on the natural flow of conversation and exchange of information.
2. The associated quantifier is a proportion, rather than a mean opinion score.

Interpretation

Since the possible causes of effects reflected in P[UDI] include impairments that degrade the usability of a connection, low MOS values may

TABLE 5-3

Criteria for Connection Usability Defined for Test Subjects in Service Attribute Tests

Reported Effect	Description
Unusable (U)	Connection was so poor that the only viable option would have been to hang up and try again, if you weren't paying me to stay on the call.
Difficult (D)	One party or the other had to ask for repetition of information, speak louder to be heard, slow down and articulate more clearly, or otherwise adopt an unnatural conversational pattern in order to maintain the conversation.
Irritating (I)	No actual difficulties were noticed, but the impairments or problems were great enough to represent a distraction or an irritant.
Noticeable (N)	Some impairments or problems were noticeable, but were neither great enough nor persistent enough to materially affect the conversation.
None (O)	Either no problems were noticed, or they were so slight or sporadic as not to represent any meaningful effect on the conversation.

correlate with P[UDI]. However, it is possible to have a high MOS, indicating satisfaction with what is being heard, and still have a high P[UDI], indicating user dissatisfaction with what is experienced. For example, suppose 1 call in every 10 placed by a user had a very loud echo, rendering the call difficult or unusable, while the remaining 9 were completely unimpaired. Then the five-point MOS would be something like 4.05. That 4.05 would be presumed to indicate acceptable quality of service, but it is unlikely that many users would be satisfied with it. This example shows that MOS calculations can mask possible reasons for user dissatisfaction that are revealed by the P[UDI].

The other major difference between the P[UDI] and a mean opinion score is that P[UDI] as defined by Table 5-3 reflects effects of problems that do not affect the quality of what is heard over a voice connection. Principal among these effects is the way that excessive round-trip delays can disrupt normal conversational rhythms, which is in no way dependent on any of the other manifestations reflected in a measure of connection quality. This means that it is entirely possible for end-to-end connections to sound just fine, but still be unsatisfactory to users. This possibility was clearly exhibited in the early experience with voice services carried over satellites. Relative to their terrestrial counterparts at the time, digital satellite links were quiet, clear, and virtually lossless. When they were protected with echo cancellation, they were also largely echo-free. Even so, they were found to be unsatisfactory by many users, in part because the greater round-trip delays increased the severity of echo when it did occur, but more often because the additional 0.5-s round-trip delay disrupted normal conversational rhythms.

As described in Chap. 4, the possible effects of packet switching on perceived quality of service include exacerbation of echo and disruptions of conversational rhythms due to greater round-trip delays like those experienced with voice over satellite. It will therefore be very important that any scheme for evaluation of likely user perception of quality of packet-switched voice services be based on both the MOS and an analog of P[UDI].

Measurement and Evaluation Tools

Quantification of MOS and P[UDI] in a way that assures these measures will serve as reliable indicators of likely user perception of voice quality over telecommunications connections is accomplished either by subjective testing or application of models of user perception of acoustic signals.

Subjective Testing

By far the oldest and most reliable medium for quantifying measures of perceived quality of voice services is subjective testing, in which user assessments of quality are elicited and collected directly from typical users of a service. Because the responses come from persons who will be using the service, and have compared it with the familiar domestic telephone service, such direct subjective testing is a very credible vehicle for evaluating quality.

Despite the intuitive appeal of gathering opinion scores and assessments of effect from which MOS and measures like P[UDI] can be calculated directly, subjective user tests are sometimes viewed as something to be avoided, because they are perceived as being unreliable, uninformative, or too demanding. There is, for example, always the fear or suspicion that the set of test subjects involved may not fairly represent the tastes, dislikes, and sensitivities of the population as a whole. Moreover, as described in the earlier discussion of interpretation of MOS, the results of any subjective test may be as much a reflection of the way the testing was conducted as of the experience of the test participants. This means that subjective testing must be carefully structured, controlled, and standardized for a particular application, creating daunting test requirements like those shown in ITU P.800. Finally, one of the major reservations is that after all the effort necessary to assure that the test results will be credible and meaningful, the data on user assessments may be uninformative. A well-structured subjective test may, for example, produce results that reliably indicate that users will not find the tested service to be satisfactory, yet reveal little or nothing about what might be done to improve it.

Notwithstanding these reservations about subjective testing, experience has shown that carefully designed subjective user tests can be a very cost-effective medium for obtaining data on perceived voice quality. Such a test can, for example, be set up in the operational environment and executed with a fraction of the time and effort that might be needed to instrument a voice service for quality testing. In addition, when designed to do so, a subjective test can quickly yield very clear answers to such questions as:

- Will the quality of a particular voice service be acceptable to its users?
- If not, why not?
- What system-level performance factors are degrading voice quality?

To assure that a subjective user test can become this kind of tool for measurement and evaluation of service quality, however, the test must incorporate four essential elements: operational realism, efficient sampling, a basis for comparison, and test caller orientation.

Operational Realism The test interfaces, environmental conditions, and activities should replicate those that will be encountered by the intended users of the service, in the envisioned operating environment. In practical terms this means that the testing should:

- Take place in the users' operational environment, or in an emulation of that environment that replicates the equipment, physical surroundings, ambient noise, and connection characteristics expected where the service is to be used
- Involve test calls in which actual conversations are evaluated

Efficient Sampling Subjective tests in the operational environment are often designed like surveys, to elicit evaluation of one or two calls from a large number of test subjects in order to achieve statistical confidence in the results. As long as answering the questions previously listed is the objective, the same ends can be achieved much more efficiently by using a relatively small number of test callers, who make repeated calls. The trick is to ensure that the test calling takes place under a schedule to create samples in which each test caller places:

- The same number of calls to every destination in the test
- A large enough number of repeated calls to each destination to support statistically meaningful comparisons

Because it enables the test designer to set up cooperating stations from which calls are originated and terminated, this kind of sampling is very readily scheduled and executed, enabling production of meaningful results in a very short time.

A Basis for Comparison As suggested in the earlier discussion of possible pitfalls in interpreting mean opinion scores without some point of familiar reference, it may be impossible to ascribe meaning to the assessments of a group of test subjects. It is therefore critical to assure that the test results can be meaningfully interpreted by including in any subjective test samples of calls completed via a *reference service*. Such a reference service is something like today's public switched telephone network with which all test subjects will be familiar. Inclusion of test calls placed via the selected reference service to the same destinations as a basis for comparison assures that the assessments reported by the test subjects can be correlated with those from a service that is known to support acceptable voice quality. The test subjects' evaluations of that known service also support calibration of the test results from a small population against the total population of users of voice services, thereby enhancing the efficacy of the efficient sampling techniques just described.

Test Caller Orientation Another feature that enhances the efficacy of the efficient sampling techniques is a reduction in ambiguity of test caller reporting. This can be accomplished by training test callers before the testing, to make sure that all involved in the tests clearly understand how to formulate and report their evaluations of the quality and usability of each test call. The test caller orientation of this kind should, for example, include instruction on:

- Placement of calls in accordance with the schedule and the conversational requirement for each test call

- Discrimination and reporting of severity of impairments experienced on each call
- Grading and reporting connection quality based on what was heard
- Grading of connection usability based on what was experienced

A subjective test incorporating these four elements results in something that is simple to set up, easy to implement, and can be conducted over a short time with relatively few people. At the same time, such a test quickly produces results that are intuitively credible, operationally meaningful, and scientifically defensible. For example, in conducting service attribute tests designed to these standards it was possible to produce unassailable comparisons of service quality among the major long-distance carriers, together with clear reasons for any failures to achieve acceptable quality. The necessary testing was often arranged and set up in a matter of days, utilizing as few as nine test callers calling prearranged destinations, and completed with production of actionable results in less than a week.

User-Perception Models

A carefully designed subjective user test, then, is the recommended vehicle for direct quantification of measures of voice quality. The alternative is indirect quantification of measures of connection quality and usability via models that generate MOS and P[UDI]-like measures of quality from objective measures of characteristics of a voice connection that can be quantified with test devices. Such models are preferred over subjective tests, because the input data can be readily collected automatically, interpreted without having to deal with the vagaries of human opinion, and replicated in different environments without worrying about typicality of the data collected.

All models of this kind are based on quantitative descriptions of user reactions to various kinds of deviations from natural speech. As illustrated in Chaps. 7 and 8, however, there are two distinctly different types of such models, broadly distinguished by whether the deviations from natural speech are expressed in terms that are:

- *Psychoacoustic.* Describing effects derived from studies of the physiology of human hearing and psychology of speech cognition.
- *Electroacoustic.* Describing characteristics of the analog and digital waveforms that result when human speech is turned into

electrical signals, encoded, transmitted, decoded, and received at the distant end of a voice connection.

Although the functions involved may appear to be similar, there is a subtle difference between these two types of models. The effects on perception of voice quality reflected in psychoacoustic models depend on the particular speech waveform affected. Thus, for example, the impact on quality ascribed to the effects of an impulse noise burst on a phoneme will depend on whether the phoneme affected is a vowel sound, a plosive consonant, or a fricative consonant. In contrast, the effects on perception of voice quality captured in models based on electroacoustics will be expressed as functions of parameters, such as signal power and background noise level, whose values are independent of the semantic content of the waveform affected.

The necessity of distinguishing the semantic content of speech waveforms in applying psychoacoustic models creates a practical limitation on how connections can be tested in order to acquire data needed to apply the model. In the case of electroacoustic models, the necessary parameters may be objective measures of gross end-to-end connection characteristics that can be quantified with any number of test devices and test waveforms. In the case of psychoacoustic models, however, the objective measures must be derived from the comparisons of transmitted and received versions of test waveforms with a known phonetic content. This means that the test must involve transmission and reception of carefully constructed test signals, or segment-by-segment processing of captured versions of speech waveforms to identify the likely phonetic content of the signal.

The problems posed by acquiring data for use with a psychoacoustic model may, then, be the deciding factor in the choice of what type of model to use in a particular application. Beyond this, however, the selection of which of possibly many competing user-perception models should be used for a particular purpose is likely to be a daunting proposition. It is a good bet that at almost any point in time there will be a variety of models of either type, being proffered by advocates and endorsements as *the* solution to the problem of measuring and evaluating voice quality. The problem in the ongoing debates as to which should be used will be that there will probably be no absolute basis for deciding which of two models gives the "right" results. For example, suppose we are interested in determining whether the quality of an untested voice service will be acceptable. We gather test data for, and apply, two models. The first model, A, predicts a MOS of 4.5, while the second model, B, predicts a

MOS of 4.1 from the same test data. Then conventional wisdom would have us believe that one of those models must be right, while the other must be wrong, because the difference is too great to be explained by sampling variance. However, as suggested in the earlier discussion of MOS, it is entirely possible for both answers to be right, as would happen, for example, when application of model A to familiar domestic long-distance service produces a MOS of 4.3, while application of model B to that service produces a 3.8.

This example shows that the decision to adopt one model or another must ultimately rest not on some notion of what is the universal truth of the matter, but on the answer to the question: Is the model useful for its intended application?

In the case of user-perception models, the principal function of the model is to enable us to predict likely user perception of voice quality without having to elicit subjective evaluation from users of the service. The ultimate *application* of those predictions will be to produce an answer to some question about an untested voice service, such as:

- Will the quality of a service be acceptable to users?
- If not, why not? What needs to be changed to make it acceptable?
- What system-level performance factors are degrading voice quality?

The principal criterion of assessing the utility of a user-perception model is, then, the model's *reliability,* as gauged by the probability that the answer to the question will be correct. Since the answers to these questions are qualitative, the question of reliability is not one of accuracy of whatever numbers are produced by the model, but whether the *interpretation* of those numbers produces correct answers to the target question. In practical terms, this implies that the only real test of reliability of a user-perception model is the demonstration of its success in producing correct answers.

There are, however, some criteria that can be used to gauge the likely reliability of a user-perception model when there is no history of its success in the intended application. The most common of these are tests for completeness, consistency, and correlation.

Completeness The test for completeness of a model relative to its intended application is whether all effects that can degrade likely user perception of voice quality which might be manifested in the service being evaluated are reflected in the model. More formally, this criterion simply states that when users will be significantly exposed to impairment X in a particular voice service, then the underlying model must have some independent variable x that is associated with impairment X and some dependent variable whose value is affected by variations in x. This condition is very easy to verify from a description of model inputs and outputs, yet has sometimes been completely overlooked in the heat of debate of the relative merits of competing user-perception models for some purpose. Application of this criterion would have readily shown, for example, that a psychoacoustic model utilizing results obtained by transmission and reception of phonetically structured test signals would never be an adequate model for voice services in which there is a possibility of echo, because the test data would never be affected by echo path delay.

Consistency Another easily effected test for likely reliability of a user-perception model is straightforward verification that the model predictions exhibit both external and internal consistency with our expectations. External consistency is readily verified simply by ascertaining that the model predictions based on parameters for known acceptable services indicate that the service will be acceptable, and the model predictions for extremes of poor quality indicate that the service will be unacceptable. Internal consistency is verified by testing sets of input values to verify that sequences of combinations of values that should predict steadily decreasing user perception of quality actually do so. Thus, for example, a test for internal consistency of a user-perception model based on circuit loss and noise might involve examination of some test cases in which combinations of loss and noise values are ordered by the severity of the expected impact on user perception of quality and the corresponding model results are tested for preservation of that order. This kind of internal consistency is so obviously essential to achieving model reliability that it is often induced by the formulation of the model.

Correlation By far the most widely used evidence of the reliability of a particular user-perception model is application of the model to the results from controlled subjective tests to verify that the quality indicators produced from the model exhibit a high statistical correlation with the subjective assessments of quality (and/or usability) elicited from the test participants. The reasoning in this case, which is sometimes not articulated, is that a high correlation between subjective test responses and the model predictions demonstrates that the model is both complete and internally consistent and has also accurately captured the mechanisms that shape user perception. The reasoning here is valid so far as it goes. However, there are two caveats that must be kept in mind when using correlation studies to assess the utility of a user-perception model:

1. *Statistical correlation only describes the concomitance of the* averages *of the predicted and actual values.* This is a mathematical truism that need not be proven here. What it implies in practical terms, however, is that it is possible to have a high correlation between the actual and observed mean opinion scores that does not substantiate the predictive quality of a user-perception model. The breakdown here can occur, because, for example, the average of the predicted MOS values was 4.0 and corresponded very well to the MOS of the elicited responses for that particular test case, but the average value of 4.0 was obtained from a set of predicted values that ranged all the way from 1.0 to 5.0, while the testers' opinion scores only ranged between 3.5 and 4.5. In this case, the test for correlation masked a weakness in the way the model produced predictions that would be very detrimental to some applications, such as the use of the model predictions as the basis for deciding when service quality had deteriorated.

2. *The model examined may have been calibrated by such correlation studies.* Without a detailed examination of the way that a particular user-perception model was derived, it is impossible to know whether the high correlation between a model output and a subjective test result was a consequence of the accuracy of the model or the way that the model was constructed. There are, for example, some user-perception models that are constructed in such a way as to allow the user to calibrate the model for a particular environment by maximizing the correlation between outputs and inputs to the model. Moreover, nearly every user-perception model is developed empirically, to achieve agreement between the model outputs and some body of subjective test results. This by itself is not a limitation. However, as described earlier, the actual values of MOS or other indicator of user perception of quality, will depend on the structure of the subjective test that produced those results. Thus, the correlation expected will depend critically on the question of what variety of testing was used. For example, a high correlation between the values of MOS predicted by a model and those elicited in a listening test conducted in accordance with the guidance in ITU P.800 to validate the model may imply absolutely nothing about the predictive value of that model for an operational telephone service.

Beyond these criteria, which comprise part of the basis for determining whether a model is *scientifically defensible,* the utility of a user-perception model for some purposes will depend on inherent credibility, extensibility, and manipulability of the model.

Credibility As indicated earlier, the ultimate application of a user-perception model will be to answer some question whose answer depends on how users will react to the performance characteristics of a particular voice service. Before comfortably accepting that answer, the decision maker who posed the question must have confidence in the model results, either because of known successful applications, or because the model itself is perceived by the decision maker to be logical and natural enough to be accepted at face value as a reasonable means of effecting predictions. This means that the value of a user-perception model as a decision support tool will depend to a large extent on its *inherent credibility*, determined by the ease with which the decision maker can understand the model's structure and functioning without delving into the mathematical details. When a user-perception model has high inherent credibility, because input parameters are readily recognizable as something that clearly relates to quality of a voice connection and the model's processes can be easily understood as a reasonable mathematization of real-world interactions, the decision maker can comfortably use it. When a model has low inherent credibility, because the structure is mathematically complex, arcane, or opaque, the decision maker will continually second-guess the model results, and reject its use altogether the first time that a course of action recommended as a result of interpretation of the model conflicts with the decision maker's intuition. This means that when there is a choice between one very precise model that is not inherently credible, and another, adequate model that is, the one with inherent credibility will be much more cost-effective in the long run.

Extensibility Another factor affecting long-term cost-effectiveness of the user-perception model selected for a particular application is the ease with which the model can be extended to work equally well in other environments or other applications. The inherent value of such extensibility in any model is so obvious that it may hardly seem worth the mention here. Yet, it is clear that user-perception models have in the past been evaluated and endorsed without any consideration of extensibility. For example, the PESQ (Perceptual Evaluation of Speech Quality) model is endorsed in ITU Recommendation P.862 (Ref. 5) as "...an objective method for predicting the subjective quality of 3.1 kHz (narrow-band) handset telephony and narrow-band speech codecs." The model is endorsed therein as a means of answering the question, "How good will speech carried via a handset over a particular codec sound?" It was later endorsed as the preferred model for gauging fidelity of speech signals exposed to dropped frames during transport via a VoIP service. At the same time, ITU P.862, paragraph 4, declares, in effect, that the endorsed model is definitely not extensible, by stating that:

> It should also be noted that the PESQ algorithm does not provide a comprehensive evaluation of transmission quality. It only measures the effects of one-way speech distortion and noise on speech quality. The effects of loudness loss, delay, sidetone, echo, and other impairments related to two-way interaction (e.g., center clipper) are not reflected in the PESQ scores. Therefore, it is possible to have high PESQ scores, yet poor quality of the connection overall.

This quotation demonstrates the sense in which a model, in this case the PESQ, may not be extensible with respect to applicability. Other facets of extensibility to weigh in the assessment of the long-term utility of a model are the ease with which it can be extended to incorporate the new effects or accommodate new data. A user-perception model that was extensible with respect to incorporation of new effects would, for example, be one that was an adequate predictor of user perception of voice quality in the face of all the familiar impairments in telephonic voice services, but could be readily augmented to reflect the effects of something like dropped frames, which do not occur in circuit-switched services. This kind of extensibility is particularly valuable, because the extended model capitalizes on all the experience, refinement, and credibility achieved in the earlier version instead of starting from scratch.

A model that is readily extensible with respect to new data is one in which any factors that are assumed or derived from empirical data are represented as parameters in the model that can be readily changed. An example of this kind of extensibility in a user-perception model would be reference points in the AT&T loss/noise grade of service model shown in Table 5-4, together with PGoB, the percentage of calls rated "good" or better for those loss and noise values in two different subjective tests. As suggested by Table 5-4 and described in App. C, the model can be readily recalibrated for different environments by conducting limited subjective tests of perception of voice quality over connections controlled to exhibit those reference values to revise, or assume new values of, the associated PGoB scores. As shown by the differences in measures in Table 5-5, the reference points can similarly be used in conjunction with known correlations in values of different measures to create other models using different parameters. These examples immediately suggest, then, that like inherent credibility, extensibility of a model enhances its long-term cost-effectiveness—the greater the extensibility, the less the overall cost of achieving a means of answering different questions in different environments.

Manipulability By manipulability of a model is meant the ease with which inputs can be varied to achieve different outputs. The obvious benefit is that when a model is openly manipulable, it can be used to go beyond simple prediction to an examination of tradeoffs and exploration of what-if questions. Although such manipulability is sometimes considered to be the sine qua non of a useful model, many of the user-perception models proposed or endorsed over the years have not been manipulable at all. The most obvious examples are those psychoacoustic models whose application is based on comparison of transmitted and received versions of specially constructed speech waveforms. Processing of samples of connection acquired by recording the received waveforms produces estimates of MOS that are unquestionably reliable for some applications. However, there is no mechanism whatsoever by which that data or the results can be examined to answer the question, "What would have happened had the [name(s) of objective measures of connection characteristics] been different?" This is not to say that such psychoacoustic models are not useful in their proper application. Rather, it simply shows that such models are patently limited, capable of revealing information only for the cases for which samples were collected, and

TABLE 5-4 Reference Points Established for the Original AT&T Loss/Noise Grade of Service Model

Transmission Signal Loss, dB	Circuit Noise, dBrnC	Transmission Rating Factor *R*	PGoB Test 1, %	PGoB Test 2, %
15	25	80	86.4	63.9
30	40	40	6.5	0

TABLE 5-5 Reference Points Established for the *CCITT Blue Book,* 1989, Loss/Noise Grade of Service Model

Overall Loudness Rating, dB	Circuit Noise, dBmp	Transmission Rating Factor *R*
16	−61	80
31	−76	40

therefore would be of lesser value for decision support than an equally reliable model that was manipulable.

Summary and Preview

The purpose of this chapter has been to lay the groundwork for addressing the central problem of assessing likely user perception of the quality of voice carried over packet-switched networks. Chapters 6 to 8 comprise a survey and assessment of the state of the art in voice quality measurement and evaluation technology. In those chapters the broad principles of evaluation developed here will be applied to test protocols and models currently in the marketplace to support recommendations as to which techniques offer the greatest utility and value.

CHAPTER 6

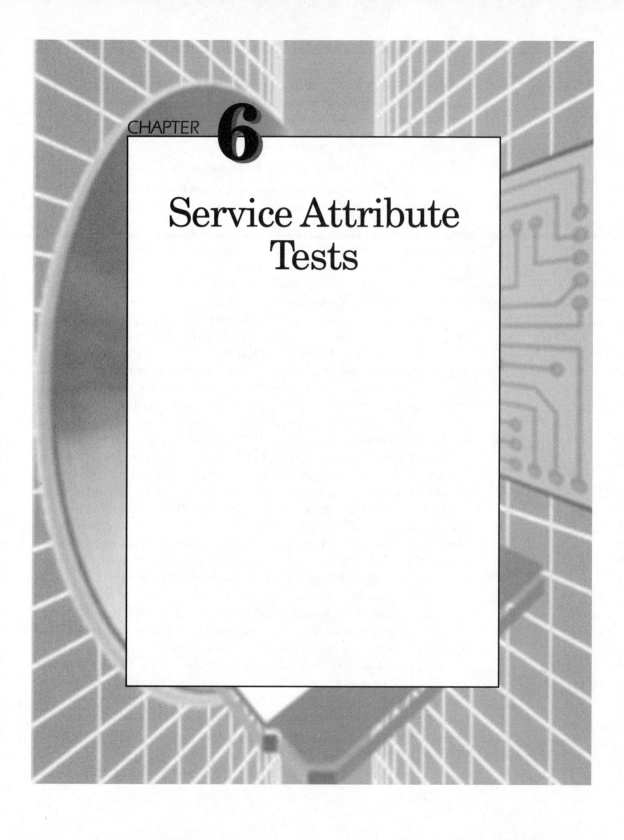

Service Attribute Tests

In Chap. 5 it was observed that one of the most versatile tools for measurement of likely user perception of voice quality is a carefully constructed subjective user test, whose design

- Achieves *operational realism*
- Employs *efficient sampling* strategies and *tester orientation*
- Produces a *basis of comparison* of both connection quality and connection utility of the service tested with a familiar service used by the testers

To illustrate the utility of such a tool, this chapter describes a subjective test protocol, shows how to structure its application to achieve these characteristics, and shows how those design features produce results that can be used for operational decision making.

The particular protocol chosen to illustrate design criteria for subjective user tests is the *service attribute test* (SAT). This test was developed by Satellite Business Systems in the early 1980s, because subjective tests in use at the time were found to be inadequate for purposes of analyzing long-distance voice services employing satellite transport. Early in the deployment of its commercial satellite voice services, the company was encountering situations in which users were reporting that the satellite voice service sounded really loud and clear, but were still complaining about its quality. Attempts to use then-standard subjective test protocols to try to determine why a voice service that was lossless and virtually noise-free was not acceptable to users pointed up the necessity to expand subjective testing in two ways. The then-standard tests were showing user dissatisfaction, but providing little insight as to why users were rating the connection quality lower when it was expected to be higher. This demonstrated the requirement for designing subjective tests that would consistently elicit information as to what the users were reacting to. In addition what information *was* being derived from free-form comments was suggesting that some of the dissatisfaction was due to the greater transmission delay over the satellite, which had little to do with how clear the voice was. This suggested the necessity to include user assessments of what we have characterized as connection usability, as well as the assessments of connection quality that were being collected. As described here, the resultant subjective test tool achieved the kind of inherent credibility, extensibility, and manipulability of results cited at the end of Chap. 5 as features that enhance the utility of user-perception models.

Basic Test Structure

The service attribute test is a *conversational* test conducted by a relatively small number of participants. The persons who will be reporting their assessments of quality are referred to here as *test subjects* to distinguish them from other participants. They must be ordinary users of the telephone and *not* conversant with the technology or engineering of telephone systems. They use ordinary handsets and business lines to place calls to preselected stations answered by cooperating test participants, who may also be test subjects, or simply persons paid to answer the calls and hold conversations. A test call is executed by dialing the number of a participating station, engaging in a brief conversation when the distant station answers, and recording specific data elements that describe the experience with that call attempt. Any particular routing of test calls is set by the number dialed and/or the handset from which the call was placed, assuring that test subjects have no knowledge of the path taken by any particular call.

Data on Calls

The purpose of the test is to produce a body of reports from each of the test subjects showing for each call placed:

- Identification of the call
- Outcome of call attempts
- Presence and severity of impairments noted
- Assessment of the overall quality of the connection
- Description of effects of impairments and other connection characteristics on usability of the connection

The data form may also include places for recording answers to questions of call quality elicited from called parties that are not test subjects, and a field for free-form comments or a description of strange sounds heard, particular difficulties experienced, etc. Before the actual test calling, all test subjects undergo training in which they are instructed to report these items.

Call Identification

In addition to the number dialed, data identifying each call comprises the following.

Identification of the Test Subject Placing the Call For purposes of data quality control, it is essential to know which test subject evaluated each call. Each data-collection form therefore has a place for full-name identification of the tester, which is later recorded in the test data base in a record field containing unambiguous caller initials.

Handset Used for the Call Inclusion of an identifier for the handset is another essential element, both for data quality control and for test setup. Identification of the specific handset used for a call is necessary for detection and elimination of possible bias in the results due to handset defects. Accordingly, each handset used in the test is labeled with a simple identifier to be recorded for each call or set of calls placed with it. This device also enables blind testing of different services. To avoid using different numbers for routing calls via different services, which might provide a clue as to which service is being used, PBXs or terminating switches can be programmed to selectively route calls originating from particular handsets. The testing of different services can be effected by having the testers dial the same number for any particular destination, and move around among different sets of stations (handsets).

Time of the First Call Attempt Because SATs are usually executed on a tight schedule, test callers are required to carefully monitor the time at which each call is placed. Such attention to time is encouraged by having the test callers begin the data recording for each call by writing in the number to be called and recording the exact time dialing began.

Outcome of Failed Call Attempts

Although the test calls are placed to cooperating stations that are supposed to answer, it is possible to encounter a number of conditions that will prevent a call from being answered the first time it is dialed. To ensure that a scheduled test call cannot, in fact, be completed, test callers are instructed to make five attempts before concluding that the call is not going to be answered. They are instructed to recognize and report the reason for each failed call attempt as being:

- *RNA.* No answer after 10 rings
- *SBY.* Slow busy signal, indicating that the called station is busy
- *FBY.* Fast busy signal, indicating that the call cannot be connected due to congestion
- *RVA/SIT.* A recorded voice announcement of the inability to connect the call, which may or may not be preceded by the warbling special information tone
- *H/D.* The call attempt went "high-and-dry," resulting in no response at all

The data on completion problems produced in this way quickly reveal any abnormal problems in completing calls, so that the necessary corrections can be made in time to avoid loss of quality data. The data also support calculation of the *normal completion rates,* calculated as the proportion of call attempts resulting in an answer, ring-no-answer condition, or a slow (station) busy signal. The normal completion rate defined this way is a gauge of likely user perception of quality with respect to connection reliability. It can be analyzed to surface possible connectivity problems, demonstrate differences in performance between two different services, etc. (see Ref. 1, Chapter 8).

Impairments Noted

When a test call is connected, the test subject is expected to hold a brief conversation with the distant party, talking about anything of mutual interest *except the quality of the call or what is being heard.* The restriction on talking about quality does not, however, preclude the called party from *commenting* on unusual phenomena difficulties when that person is not a test subject. Moreover, when the called party is not a test subject, the test subject reporting the call is instructed to end each conversation by soliciting and recording answers to two classical subjective user test questions:

- Did you have any difficulty talking or hearing on this call?
- How would you rate the quality of this call?

The test subjects are instructed to hang up the phone, and only then report their assessments of what they heard and experienced in the required formats.

The first element of that reporting is a rating of the presence and severity of impairments to the quality of what was heard over the connection. The types of impairments to be listened for and reported may vary. However they will be set for any particular SAT before the testing begins by defining and describing for the test subjects readily recognizable manifestations of conditions expected to degrade the quality of the connections sampled. The principal impairments that have been defined for nearly every SAT that has ever been conducted are

- *Low volume.* Power of the speech signal received over a telephone connection is noticeably lower than what would be expected in a face-to-face conversation. ("The distant speaker sounds far away, is hard to hear, sounds too soft, is muffled, etc.")
- *(Constant) noise.* Audible background noise that is present while the distant taker is silent or that is left over after the ear has filtered the signal and extracted the emulated natural speech. ("There is persistent static, crackling, hum, or buzzing on the line.")
- *Speech distortion.* Qualities or characteristics in the received speech signal that would never be heard in face-to-face conversations. ("The speech sounds raspy, muddy, or wispy. Sounds like the distant speaker is talking underwater or gargling. Sounds strange, weird, or warbly.")
- *Echo.* Manifested exactly the same way as echo in an empty stadium or a canyon, as a return to the ear of something spoken and delayed long enough to be recognized as speech. ("I can hear myself when I talk…talk.")

Other impairments that have been distinguished for SAT subjects from time to time because they were likely to occur with significant frequency in the particular services being tested include:

- *High volume.* Perceived when the power of the speech signal received over a telephone connection is noticeably, perhaps even uncomfortably, louder than what would be expected in a face-to-face conversation. ("I'm being blasted by the distant speaker. Too loud!")
- *Crosstalk.* Presence of audible conversation that is being carried on some other connection. ("I can hear someone else talking on this line.")
- *Impulse noise.* Manifested as intermittent, very short, high-power noise surges on the line. ("There is popping on the line.")
- *Noise on speech.* Perceived when there is a distinct difference between the noise on the line when the distant parties are speaking and when they are not.

- *Incomplete words.* Perceived as gaps in signal power producing missing phonemes or syllables in words recognized in the emulated natural speech signal. ("It sou-s li- some-ing is missing.")
- *Clipped words.* Perceived as gaps in signal power producing missing phonemes in the beginning or endings of words, but not in the middle. ("It -ound- li- -omthin- is -issin-.")
- *Garbling.* Manifested as complete loss of intelligibility in the emulated speech signal, even though the speech signal can still be discriminated by the human listener. ("I cn t ll wu s bng sd.")

As suggested by these examples, the particular impairments defined and described in a SAT can be almost anything that is, in the terms introduced earlier, a readily recognizable transmission artifact that can be clearly and unambiguously described for the test subjects. The important thing in their selection is that they represent something that is expected to occur in the services being tested and are described in such a way as to pose no problems for training test subjects in their recognition.

In reporting the incidence and severity of such impairments, the test subjects are trained to recognize each impairment and rate its presence in each call as being:

- *None.* The impairment was not present or was so slight as to have a negligible effect on quality.
- *Much.* The impairment was definitely present and noticeable enough to degrade the quality of the call.
- *Some.* The impairment was noticeable, but sporadic, or otherwise not severe enough to be described as "much."

Under this rating scheme, the definitive descriptions are "none," which is easy enough to assign, and "much," which is similarly easy to assign, because it represents conditions under which an impairment was definitely present and definitely an impediment to listening. The "some" category is included to allow the test subjects to comfortably report the presence and severity of impairments that were clearly present, but could not fairly be described as an impediment to listening. The three-level reporting scheme thus creates a "fuzzy" spectrum like that shown in Fig. 6-1, in which progressively greater deterioration with respect to a particular impairment will smoothly transition from a state in which users are nearly certain to assign "none," through increasing probabilities of assignment of "some," and thence through decreasing probabilities of assignment of "some," to a certainty of assigning "much."

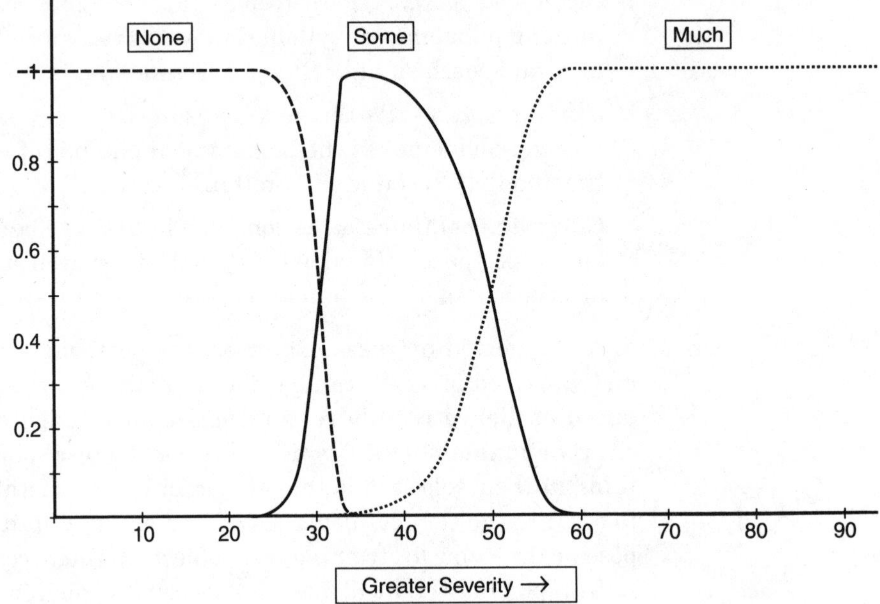

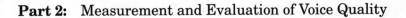

Figure 6-1 Probability of reporting impairment severity as "none," "some," or "much" as severity increases.

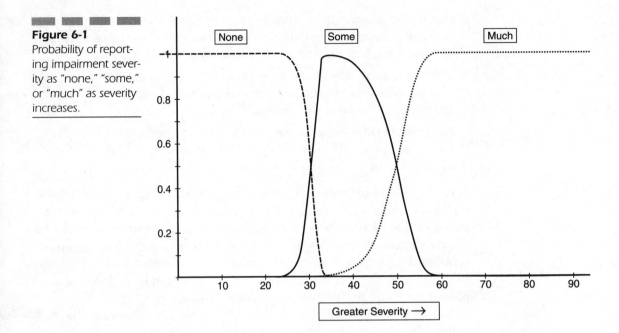

Assessment of Overall Connection Quality

After reporting the incidence and severity of the impairments specified for the SAT, the test subjects are asked to rate the overall connection quality as perceived at their side of the connection. When the called party is not a test subject, the test subjects also ask the distant party for an assessment of call quality. The subjective ratings are selected from a list of five qualitative descriptors:

- Excellent
- Good
- Fair
- Poor
- Unusable

The scale of numerical opinion scores traditionally used in SATs is the 0 to 4 scale from the late 1970s, in which 4 denotes "excellent," 3 "good," and so on, with "unusable" being assigned 0. The SAT instructions also explicitly allow the test subjects to employ half-point scoring for shades of difference, so that reported scores of 3.5, 2.5, etc., are valid. When the persons called are not test subjects, however, the description of quality is supposed to be forced. This means that half-point scores are not valid unless the called party's rating of connection quality is a waffled qualitative answer like "very good" or "good-to-fair."

Description of Effects

One of the distinguishing features of a SAT is that the test subjects are trained to distinguish and report the effects of what was experienced on the usability of the call. The description of those effects is reported by assignment for each call of one of the following grades:

- *O—no effects.* No impairments degraded the quality of the connection or impeded its use for conversation.
- *N—noticeable.* Impairments were noticeable, but did not impede the usability of the connection.
- *I—irritating.* The conditions on the connection did not cause difficulties in the sense of changing normal conversational patterns, but what was experienced was something that was or would become a definite irritant to the user.
- *D—difficult.* Impairments or other conditions on the connection materially disrupted the normal flow of conversation, for example, by forcing the call parties to raise their voices, ask for repetition of information, or adjust to abnormally long hesitation in responses to questions. Such difficulties would also include situations in which the call parties encountered unwarranted instances of both parties talking at once, or had to wait longer than usual before deciding whether to respond to the distant party.
- *U—unusable.* The conditions on the call were so severe that the test subject or the called party would have hung up and redialed the call had that been an option (which it isn't because the SAT participants are being paid to place the call and hold a conversation, no matter what happens).

In this scheme of grading, the unambiguous responses are "O" (no effects), which is easy to assign; "D" (difficult), for which the criterion is the experience of some recognized difficulty during the course of the call; and "U" (unusable), for which the extent of the difficulties is great enough to create the impulse to hang up. The options of reporting "N" (noticeable) and "I" (irritating) are much more subjective. They are included to allow for comfortable descriptions of shades of differences, serving the same purpose as the "some" category for description of impairments. The inclusion of two transitional categories in this case also allows the test subject to grade the experience as being more like that with a good call (noticeable) or a bad one (irritating).

As described earlier, the measure of quality derived from this aspect of the SAT data collection is P[UDI], representing the proportion of evaluated calls for which the overall effect of everything that was experienced on that call was deemed by the test subject to have rendered the call "unusable," "difficult," or "irritating." This measure reflects the combined impact of both what was heard and other conditions, such as excessive delay between articulation of an idea and hearing the expected response from the distant party. It therefore serves as a good indicator of likely user perception of connection usability.

When the called party is not a test subject, the analogous assessment of connection usability is elicited by the question, "Did you have any difficulty talking or hearing over this call?" The forced response is "yes" or "no," and the derived indicator of connection usability is the proportion of calls for which the answer was "yes." In the early days of the SAT, the inclusion of part of the data collection enabled us to correlate the SAT results with those from some of the more traditional subjective user tests in use at the time. The reporting has been retained, because the question frequently elicits unsolicited comments as to the nature or cause of the difficulty, and such comments have often been found to be invaluable in diagnosing problems with particular routes.

Features of the SAT

As indicated earlier, the structure of the SAT imbues it, as a test, with the same kind of inherent credibility, extensibility, and manipulability of results that are desirable features in a subjective perception model. These features are briefly described here both to suggest the utility of the SAT and to further illustrate the concepts.

Inherent Credibility

Extensive experience with its application shows that results from a SAT are readily believed and trusted as a basis for decision making. The principal reasons the SAT has such high inherent credibility are that

- The results describe the perceptions of ordinary users in a way that can be readily understood by almost anyone.
- The test is designed to be conducted over operational services using the same kinds of lines and telephone handsets that would be used by anyone else.
- The testing is conducted in a typical real-world environment, without any attempt to reduce the ambient background noise.
- The results reflect what users heard and experienced during the course of ordinary conversations, as perceived by persons who have no idea of what is going on over a telephone connection, *except* what is manifested in those conversations.
- The impairments can be readily understood and recognized by their descriptions.

All these features combine to reassure someone who is contemplating use of the results that the quality assessments are truly typical and not conditioned by recognition of some arcane artifacts that can only be distinguished by listening "experts" and may not even be detected by ordinary users.

In addition, examination of the reporting instructions readily demonstrates that although the SAT is still a subjective test, much of the subjectivity in the responses is minimized or controlled. Subjectivity in the assessment of quality is unavoidable, because there is no way to elicit from users the precise criteria they are using when, for example, they rate the quality of a connection "excellent," rather than "good." However, in many instances, the criteria for the other kinds of ratings, such as presence and severity of impairments, or experience of difficulty on a call are unambiguous enough to be discriminated objectively.

Because of these features almost no one for whom the structure of the SAT has been described has expressed reservations as to the validity of the methodology, results, or conclusions. Such ready acceptance the first time results from a particular test or model are presented is the hallmark of inherent credibility.

Extensibility

As suggested in the discussion of reporting of incidence and severity of impairments in a SAT, the particular impairments defined for the test subjects can at any time be tailored to omit impairments that are not likely to be encountered, or to include new ones that are unique to the environment or service being tested. For example, when circuit switching was analog, crosstalk was a relatively common occurrence. With the ever-widening deployment of digital switching and transmission, the likelihood of encountering crosstalk dwindled to the point that its incidence was negligible. When this happened, crosstalk was dropped from the list of impairments to be reported in a SAT without any effect on the utility of the test. If there were a requirement to measure and evaluate voice quality in an environment in which there was an appreciable incidence of crosstalk, however, it could be as readily added back into the test.

More generally, a SAT can be tailored for any new environment with minimal modifications. When changes in telecommunications technology occur, then one of two things happens:

- Implementation of the voice service with the new technology produces artifacts whose manifestations are distinctly different from those commonly named in the current version of the SAT.
- The change has no effect on the kinds of impairments manifested to users.

When there are no changes in manifestations, the utility of current versions of the SAT extends to the new service without change. When the difference creates artifacts whose manifestation is distinctly different from those commonly being reported, those manifestations can be unambiguously described for test subjects and added to the SAT for the new environment.

The important feature here is that neither addition nor omission impairments reported in a SAT create an incompatibility with data from other tests. For example, the data from an environment for which crosstalk was absent and therefore not reported are, in essence, simply a sample in which the reported incidence of crosstalk was always "none." Mechanically, this means that data from an older SAT in which crosstalk was reported can be merged with newer data without reports of crosstalk simply by adding an extra field, all of whose entries is "N," or whatever other token was used for a "none" response. Conceptually, it means that the two data sets are commensurate, so they can be meaningfully compared and analyzed for differences or similarities in user assessment of quality, even though crosstalk was present and reported in one environment, but not the other.

Manipulability of Results

One of the most valuable features of a SAT is that the opinion scores and assessments of effect on which measures of connection quality and usability are based are individually recorded with the description of the test subjects' perceptions of incidence and severity of impairments. The raw data from SAT therefore create a basis for determining relationships between impairments and subjective measures of quality whose value may endure long after the purpose of the SAT for evaluating a particular service has been fulfilled.

To cite but one real-world example of the value of this feature, in 1996 engineers involved in a study of echo cancellation in the MCI network came to a point that they could not complete the evaluation of tradeoffs without answering the basic question, How much does echo in the network degrade users' opinion of its quality? They were using the classical loss/noise/echo grade of service model that enabled them to estimate the degradation in percentage of calls expected to be rated "good" or better as a function of echo path loss and delay. However, what they really needed was something that would show the effects of a device that would reduce the incidence of echo in the current network, but would have no effect on other impairments. This meant that the model could not be used to answer the specific question without conducting an extensive study to determine the statistics of echo path loss and delay in the network and applying complicated statistical analysis or simulation to determine the overall effect of echo occurring with that profile.

However, it was clear that if there were a large enough body of data that would enable comparisons of user assessment of call quality (and, as we pointed out, connection *usability*) for calls placed over our network that were completed with and without echo, then the comparisons of MOS and P[UDI] would already reflect a mix of values of echo path loss and delay. A recent set of SATs comprised just such a large body of data, and a credible answer, complete with sensitivity analysis to eliminate bias due to selection of test subjects, was produced in an afternoon.

This experience was more than serendipity. It was a result of the fact that the SAT was explicitly designed with such applications in mind, to ensure that the data would support extensive manipulability of the results. Had our evaluations been based on something designed more along the lines of the subjective user tests in use at the time or some of the user-perception models that are currently in vogue, we might still have been able to evaluate the quality of our voice services, but the test data would have had virtually no utility beyond that immediate application.

Design for Effectiveness

In discussions of the possible utility of subjective user tests we have posited that the characteristics required to assure that its results will be useful and effective are operational realism, tester orientation, efficient sampling, and a basis for comparison. The first two of these are achieved in the basic design of the SAT, making it a very good medium for capturing user perception of quality and its correlation with impairments. However, much of the decision support utility of data from a SAT, or any other subjective test for that matter, will be derived from the ability to meaningfully interpret sets of test results to answer specific questions. As will be described, the expected utility of results in this role depends greatly on the latter two characteristics. These, in turn, are achieved not so much via the test protocol, but through its *data-collection plan,* which determines the distribution, characteristics, and numbers of data points to be produced from the test.

Principles of Test Design

The purpose of formal design of any test is to assure that the results obtained will both (1) produce reliable answers to whatever questions have motivated the test and (2) hold up under scrutiny or questioning of the test results. That is, we want to be able to use the data produced from the test to respond to decision makers' concerns, with respect to both *what* the data imply and the *confidence* that can be placed in those implications.

Without going into great detail about the theory of test design, suffice it to say that this is accomplished by

1. Determining the decision maker's questions
2. Deciding what measures need to be quantified and what comparisons of those measures must be made in order to answer the decision maker's questions
3. Determining the principal factors that may affect whatever is to be measured by the test

4. Selecting from those factors the ones most likely to affect the overall test results
5. Ensuring that the samples of data for classes defined by combinations of those factors can be assembled to produce an overall sample that is
   - Typical of the larger population from which it was drawn
   - Large enough to achieve high confidence levels in the estimates of the measures selected for quantification via the test
   - Sufficient for meaningful statistical testing of differences in the measured values among different subsets of the data collected for which comparisons are *required* to answer the decision makers' questions

The classical model that illustrates these steps in formal test design is the problem of conducting a national opinion poll on some question. When such a poll is commissioned, say, by a political decision maker, the objective is to determine how the population of the country stands on a particular issue. To do this the pollster will begin in step 1 by formulating an unbiased statement of the issue, perhaps phrasing it in such a way that the logical measure for assessing national opinion is the proportion of the population that is "for," "against," or "undecided." The necessary comparisons identified in step 2 then become the percentage of "for" responses as compared with the percentage of "against" responses.

Next, in step 3, the pollster will decide the major factors affecting the opinions of the population. For example, the question might be, "Are you in favor of (*favorite controversial choice*)?" The pollster will then look at all possible factors that influence an opinion on this subject, which might be almost anything, and in step 4 select what are perceived to be the major influences on the opinions of the population on this issue. These might be, for example, the state or area of the country where persons reside, sex, age, income, and/or some other number of characteristics that are seen to be likely influences on the answers to the question posed.

The way the pollster begins to take these possible differences into account in step 5, then, is to partition each factor into possibilities, use those divisions to define categories of responders to the question, and formulate a *data-collection plan* for acquiring the samples on which the poll results are to be based. The pollster's criterion for the sample to be collected is usually that the proportion of persons in each category in the sample will closely approximate that proportion in the population tested. Thus, for example, for the likely influences just named, the partitions might be each of the 50 states of the United Sates, male or female, and age intervals of 10 years for persons aged 20 to 70. By this partitioning, there are 50 × 2 × 5 = 500 categories, which would include, for example, New York/male/41 to 50 years old and California/female/21 to 30 years old. The pollster will then try to construct a sample in which the proportion of responses from 41- to 50-year-old males living in New York in the sample will be about the same as the number of 41- to 50-year-old males living in New York divided by the total number of residents of the United States.

To accomplish this, the numbers of test subjects to be included in each of the 500 categories in the sample are then defined to be integer multiples of the number of subjects in the smallest category. That integer multiple for each category will be set to represent the closest integer ratio of numbers of persons in those classes in the overall population of the United States. The total number of persons to be polled will then be determined from this distribution by selecting the smallest number of test subjects in the least populated category that will result in a total sample size large enough to support discrimination of statistically significant differences between the proportions of "for" and "against" responses. This requirement is sometimes stated as assuring that the margins of error in the poll results are small enough to support a reliable conclusion as to whether the majority of the population favors or dislikes the proposition.

When the results of the poll are published, the pollster can expect questions or challenges as to how fair the poll was, whether the poll results accurately typify the opinions of the population of the United States, or other such efforts to assess the credibility of the sample. The pollster's responses to such challenges will necessarily be based on the defense of the following:

- The polling procedure, by being able to show, prima facie, that the question was neither phrased nor asked in such a way as to encourage certain responses and that the selection of test subjects within each category was entirely random
- The size and the composition of the sample, by being able to show that it satisfies natural criteria for typicality and supports inferences that can be ascribed high confidence levels and/or low margins of error

The reason for belaboring this example is that a pollster's test design problem is a familiar illustration of what is at the heart of any effort to construct a test whose results will be both *credible* and *scientifically defensible.* Anyone who has every followed a political campaign can readily appreciate that political polls will invariably evoke:

- Charges of bias resulting from the test procedures ("The question was phrased in a way as to invite a particular response." "All the polling was concentrated in only a few cities in each state whose residences' opinions may not be typical.")
- Attacks on the reliability of the conclusions from the results of the poll ("not enough data," "too great a margin of error")
- Expressions of doubt as to the integrity of the data ("The sample is not representative of the population from which it was drawn." "A crucial factor was ignored in defining the sample.")

In light of this example, it can be surmised that the problems of test design that led to development of the formal design process described earlier are essentially ones of (1) recognizing the likely challenges to the test results, and (2) developing test procedures and a data-collection plan that will assure the ability to respond satisfactorily to the anticipated questions.

Data-Collection Plans for SATs

In the overall design of the SAT, the provisions for orientation of the test subjects and the design of the reporting schemes combine to assure that the tester responses are largely free of extremes of subjectivity. As described in what follows, other features that assure the utility of SATs can be achieved by following the formal test design process in a way that includes provisions for sampling efficiency and a basis for comparison.

Decision Makers' Questions

The basic question that motivates a subjective user test is, "What is the quality of voice over this particular service?" The design of the SAT is, however, predicated on the expectation that decision makers will probe any results of a test of voice quality with at least three other questions:

- How does this compare with some familiar service, the competition, the best we can do, or any other referent that will help me understand what the numbers mean?

- Do the results imply that we need to improve the quality to make the service acceptable to users?
- If so, what can we do to achieve the necessary improvements?

Recognition of these questions suggests that subjective user testing should involve more than the mere reporting of connection quality. In addition to producing indicators of likely user perception of quality of calls placed via the particular service being tested, the testing should also include a basis for comparison and a clear association between what users hear, experience, and report and the underlying system performance characteristics.

A Basis for Comparison The necessity to assure that a test intended to evaluate a particular service will be something against which the results for that service can be calibrated has so far been stated as if it were self-evident. This requirement can now also be defended as a logical consequence of anticipating challenges to the test results and assuring that they can be satisfactorily addressed. Any attempts to validate the results of a subjective test will invariably raise questions of the typicality of the results. Any attempt to use the test results to produce answers to specific questions like those just described will invariably raise questions of interpretation. By designing a basis for comparison into each test, we are, in essence, guaranteeing that there will be something among those test results that can be used to address those concerns. Absent such an internally generated basis for mental calibration of the test results, the only alternative for dealing with these concerns will be something like a numerical standard set outside of the test, leaving open the question of the relationships between that number and the numbers generated in the test in question. This leaves the interpretation of the measures from the test open to all the pitfalls in interpreting MOS values described in Chap. 5. The requirement for a basis for comparison can be met explicitly, by ensuring, for example, that the test includes samples of calls placed over some familiar service as well as the service being evaluated. It can also be created implicitly, by ensuring that the testing of the service being evaluated covers a sufficient variety of different origin-to-destination routes to include at least one route that exhibits the best voice quality that can be achieved. Either way, the internal basis for comparison serves the essential purpose of providing a means of transforming the data comprising the test results into information that can be used with confidence in decision making.

A Clear Association between What Users Hear, Experience, and Report and the Underlying System Performance Characteristics The likely decision makers' questions previously listed suggest that determination of quality of voice is rarely requested or tested in a vacuum. Rather, we can easily posit that prospective users of the results of subjective user tests will have the ultimate objectives of deciding whether the quality of the service tested is good enough for some purpose and if the test shows it is not, understanding what can be done to make it so. As described earlier, the anticipated necessity to address such questions was a paramount concern that was explicitly addressed in the original design of the SAT. Table 6-1 shows the association between system-level problems and impairments that may be manifested to a user. When the objective of a test is to detect particular system-level problems, these associations can be used to assure that analysis of the impairments reported by test subjects will support the test objective.

Measures and Comparisons

As described earlier, the principal measures and comparisons for assessing quality of voice will usually be values of MOS and P[UDI] that can be compared without further interpretation with commensurate values for some familiar service, or some substitute criterion for determining what the results imply vis-à-vis likely acceptability of the service. The broader questions anticipated from decision makers as they try to assimilate and act on those results, suggest, however, that requirements for other measures should be anticipated. We list some possibilities here:

1. *Distribution of reports of "none," "some," and "much" for different impairments.* The most compelling answer as to why the voice quality over a particular route or service was perceived to be inferior is often obtained from a comparison of the reported incidence and severity of impairments noted in a SAT. Such comparisons are facilitated by production of impairments matrices in the format shown in Table 6-2.

2. *Distribution of opinion scores.* Although it receives a lot of attention in the literature on evaluation of voice quality, the MOS is but one of the possible measures that might be derived from the set of opinion scores collected. In much of the early work, for example, the principal measures that were used as indicators of likely user perception of voice quality were %GoB, the percentage of calls rated "good" or better, and %PoW, the percentage rated "poor" or "worse." These measures may well come

TABLE 6-1

Illustration of Likely Causes of Impairments

Impairment Described for Test Subject	Known Possible System-Level Cause(s)
Low volume	Excessive transmission loss over analog segments Improper setting or operation of digital codec
High volume	Inadequate loss insertion on the receive side of a four-wire connection Excessive amplification at the signal origin
Constant noise	Analog line noise Inserted "comfort noise" in connections using voice activity compression to avoid deep nulls
Impulse noise	Analog impulse noise generated by a malfunctioning line card (capacitor discharge) Sporadically dropped frames in digital transport of speech signals
Noise on speech	Marked difference between inserted comfort noise and background line noise
Speech distortion	Excessively high or low analog signal levels into codecs High bit error rate on a digital transport link Dropped frames in a packet-switched digital voice transport
Echo	Excessively high transmit signal levels from origin Inadequate loss insertion on the receive side of a four-wire connection Malfunctioning echo canceller Greater than expected echo path delay on short connection routes not protected for echo cancellation
Incomplete words	Intermittent cancellation of outbound signal due to simultaneous receipt of incoming voice signal whose power level is sporadically above the level expected by the echo canceller
Clipped words	Maladjusted voice activity detector, resulting in late activation of voice transmission Lower than planned voice levels presented to voice activity detector, causing failure to recognize soft leading and following phonemes as speech
Garbling	Extreme form of incomplete words with same cause Intermittent failure of codec Intermittent failure of transmission signal (e.g., from cell phone) Simultalk "chop," occurring when two persons speak at the same time over a connection using echo suppessors
Crosstalk	Modulation of an analog loop by hysteresis from an adjacent line

TABLE 6-2

Example of an Impairments Matrix Created from SAT Data (Hypothetical Values)

	Proportion Reported as Being:		
Impairment	**None**	**Some**	**Much**
Low volume	0.901	0.088	0.012
Noise	0.953	0.045	0.002
Speech distortion	0.976	0.024	
Echo	0.951	0.033	0.016
Crosstalk	0.997	0.003	

in handy when trying to relate results of a particular test to the predictions of quality on which the design of the service was predicated. Similarly, the distribution may well clearly indicate useful information that would not be apparent from consideration of only the MOS. For example, if the distribution of opinion scores for a set yielding a MOS of 3.3 were distributed as shown for case A in Table 6-3, we would not think much of it. However, if the same MOS were derived from a distribution like that shown for case B, we would readily infer that there was some sort of a systemic problem.

3. *Normal completion rates.* Connectivity is hardly ever the principal concern when someone is looking for a test of voice quality. However, it is such a visible feature of a voice telecommunications service that the decision makers will almost reflexively ask whether the test showed anything about connectivity. This happened to us so often, in fact, that the normal completion rates calculated from SAT data were automatically added to the presentation of the results.

4. *Calibration of the test subjects.* One of the questions about a SAT that must be expected is whether the relatively small number of test subjects is sufficient to obtain opinions that would accurately reflect those of the user population. Much of the concern with this kind of typicality can be finessed in the design of the SAT, by creating internal bases for comparison. The frequency with which this question used to be raised, however, suggested the utility of a means of making sure that the subjects in a particular test did not have quality opinions that were clearly at variance with those of the larger population. To do this, we eventually created, from the large collections of SAT data we were building, a statistically stable profile of the MOS and P[UDI] that would be expected from particular combinations of reports of severity of impairments. This profile was referred to as the *theoretical SAT model* (*TSAT*). Once this was done, the

TABLE 6-3

Example of Two Distributions of Opinion Scores Exhibiting the Same MOS but Supporting Entirely Different Inferences

Opinion	Number of Reports	
Score	**Case A**	**Case B**
4.0	21	52
3.5	40	25
3.0	24	8
2.5	10	0
2.0	3	3
1.5	2	0
1.0	0	4
0.5	0	1
0	0	7
MOS	3.30	3.30

combinations of impairments reported in the overall sample could then be used to produce a predicted MOS and P[UDI] against which the test measure could be compared to assure that there were no obvious biases in reported quality due to the selection of the test subjects.

In view of these and other possibilities for measures that might be needed to facilitate interpretation or use of SAT results to answer questions beyond the basic one, we evolved the SAT report format displayed in Fig. 6-2. It may seem that the essential guidance with respect to identifying measures here is really more a matter of making sure that the data that can be used to quantify such auxiliary measures are not destroyed. However, what is more important in this context is to recognize that there are a number of measures besides the MOS and P[UDI] that may have to be tested for statistically significant differences in order to answer the decision makers' questions and assure that the data-collection plan recognizes those requirements.

Principal Factors

For the case of a SAT, the principal factors that may affect the test subjects' assessments of the connection quality and usability are fairly straightforward. They include, for example:

Figure 6-2
A typical summary of service attribute test results circa 1992.

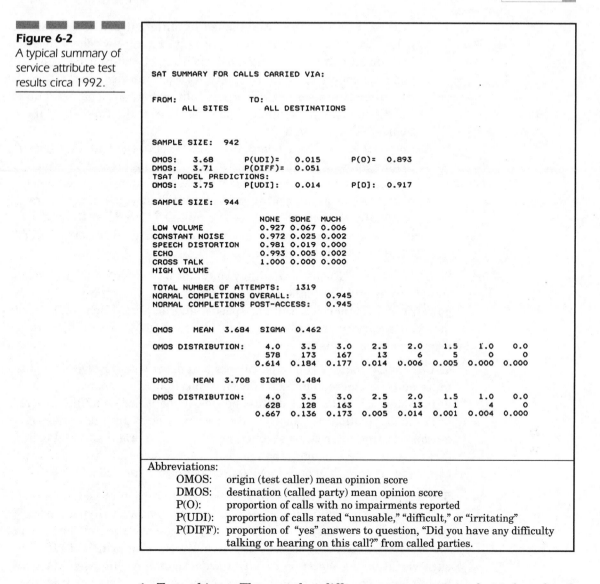

```
SAT SUMMARY FOR CALLS CARRIED VIA:

FROM:                TO:
      ALL SITES            ALL DESTINATIONS


SAMPLE SIZE:   942

OMOS:     3.68       P(UDI)=    0.015        P(O)=  0.893
DMOS:     3.71       P(DIFF)=   0.051
TSAT MODEL PREDICTIONS:
OMOS:     3.75       P[UDI]:    0.014        P[O]:  0.917

SAMPLE SIZE:   944

                      NONE  SOME  MUCH
LOW VOLUME            0.927 0.067 0.006
CONSTANT NOISE        0.972 0.025 0.002
SPEECH DISTORTION     0.981 0.019 0.000
ECHO                  0.993 0.005 0.002
CROSS TALK            1.000 0.000 0.000
HIGH VOLUME

TOTAL NUMBER OF ATTEMPTS:     1319
NORMAL COMPLETIONS OVERALL:         0.945
NORMAL COMPLETIONS POST-ACCESS:     0.945

OMOS     MEAN  3.684  SIGMA  0.462

OMOS DISTRIBUTION:     4.0     3.5     3.0     2.5     2.0     1.5     1.0     0.0
                       578     173     167      13       6       5       0       0
                     0.614   0.184   0.177   0.014   0.006   0.005   0.000   0.000

DMOS     MEAN  3.708  SIGMA  0.484

DMOS DISTRIBUTION:     4.0     3.5     3.0     2.5     2.0     1.5     1.0     0.0
                       628     128     163       5      13       1       4       0
                     0.667   0.136   0.173   0.005   0.014   0.001   0.004   0.000
```

Abbreviations:

OMOS:	origin (test caller) mean opinion score
DMOS:	destination (called party) mean opinion score
P(O):	proportion of calls with no impairments reported
P(UDI):	proportion of calls rated “unusable,” “difficult,” or “irritating”
P(DIFF):	proportion of “yes” answers to question, “Did you have any difficulty talking or hearing on this call?” from called parties.

1. *Test subject.* The way that different persons assess what is heard and experienced is the major factor that makes a subjective test subjective.

2. *Voice service.* Differences in voice services are, of course, mandatory distinctions to be preserved in a SAT. The classification of the service may, however, require discrimination of a number of different characteristics, such as:

- *Service provider.* The agent responsible for creating and managing the transport network via which the call was connected

- *Type of service.* As distinguished by different services established within the carrier network (e.g., circuit-switched voice with direct distance dialing (DDD) routing or 800-routing, packet-switched voice, virtual private network service, and actual private network service). The possibilities here are endless, but the distinctions to be made should be clear at the time any particular SAT is being defined.
- *Connection configuration.* Any variations in the way that particular connections in the service may be configured in the voice service that might affect voice quality. For example, when connections via a particular packet-switched voice service are optionally configured with G.729 and G.711 codecs, the codec used will become a major factor to be considered in setting up the data-collection plan for the test.

3. *Called party.* Even though they may not be test subjects, the voice signals from different called parties will present differences in speech power, vocal frequencies, talking speeds, and accents. Such differences may cause variations in the way that identical impairments will affect user perception of quality.

4. *Connection route.* The impairments and delay will vary with the distance and transmission path(s) of each connection. This means, at a minimum, that each call sampled must be clearly identified as falling into an origin-to-destination category. The category may be as granular as the origin and destination city, or as fined-grained as the origin site and called station, when there is more than one site from which calls are placed in a given city or more than one called station per destination city.

5. *Time of day.* Historically, time of day has not been considered a significant factor in evaluating voice quality. The reason is that the principal performance characteristic that may vary with time of day is the amount of traffic being offered to the network. The principal effect of such variation is *congestion* resulting from too many call attempts competing for the available capacity. In a circuit-switched network such congestion is manifested as call blocking, with little effect on the quality of connections that can be established. In a packet-switched network, however, traffic congestion will be manifested as a deterioration in connection quality due to greater packet latency, which affects round-trip delay, and jitter, which affects dropped frame rates through voice codecs. Consequently, the time of day may become a critical factor to take into account in the design of SATs conducted to evaluate packet-switched voice.

6. *Treatments.* Another factor that must be identified and recognized in the design of any SAT involving service(s) for which special controls have been effected for purposes of testing is the gradation of those controls. For example, when the test configuration is set up to direct call attempts through different pieces of equipment, differentiation of which type of equipment each call goes over will be critical for the credibility of the test. More challenging is the case when the test is instrumented to enable control of various performance characteristics. For such tests, it will be crucial to carefully select the combinations of settings of those characteristics that reflect points representing both of the following:

- *Expected normal operating ranges.* Selected to represent values of parameters that would be expected under normal conditions in the envisioned operating environment.
- *Extreme conditions.* Deliberately selected to push the performance sampled into regions where the incidence of consequent impairments would be reported as all "none" or all "much," as shown earlier in Fig. 6-1.

These two sets of conditions may coincide when the extreme conditions fall into normal operating ranges. However, the reason for distinguishing them is that the corresponding samples have different purposes which must be considered and recognized in the test data-collection plan. The data from normal operating ranges will be what are needed to understand what users are likely to hear and experience. The data from extreme conditions will be used to set empirical boundaries for user tolerance regarding what they hear or experience. The former characterization, then, will be what is needed to answer the principal question of likely user perception of quality. The latter had better be available to respond to the inevitable question, "How much deterioration in...can occur before users will start complaining?"

7. *Order of calling.* For other kinds of testing, the order in which data elements are acquired usually has little effect on the content of the data. However, the results of subjective user tests can be sensitive to the order in which calls are placed. The problem in this regard is that the assessments of what a test subject hears and experiences on a call will be conditioned to some extent by what was heard in immediately previous ones. For example, when there is a systemic difference in characteristics between two calls completed back to back, the contrast will be more noticeable than when the two calls are completed at different times, without an opportunity for direct comparison. This is particularly true when the two calls to the same destination are completed back to back.

Selection of Factors and Categories

The purpose of identifying the various factors that might affect the values of test measures and their appropriate gradations is twofold. The first objective is to clearly understand all the possibilities for defining cells for the test, just as our hypothetical pollster had to do before selecting state, sex, and age as the basis for the data-collection plan for the poll. As illustrated in Table 6-4 the number of cells in the test will be defined by product of categories or gradations defined for each factor and will therefore grow very rapidly with the number of factors selected. Consequently there are strong practical limits as to how many factors and classes for each factor can be selected for a particular test. It is therefore very important to take the time to identify what might be important and decide which of the myriad of possibilities is to be reflected in the test design. The best guidance that can be given for accomplishing this is to imagine that a skeptic will review the test results, anticipate the challenges, and ensure that the most likely specific questions can be answered with the factors selected.

As suggested by the last possibility described in the previous section, there is a secondary purpose to this exercise: to decide what to do about factors that are not used to define test cells. Here the options are more limited. The only alternatives for a given factor are to simply ignore it, or,

TABLE 6-4

Example of Growth of Cells for a Data-Collection Plan

	Categories			
Factor	**Definition**	**No.***	**Typical Cell**	**Cumulative Number of Cells**
Test subject	One per test subject	9	TS1	9
Network	Different providers	3	TS7/MCI	27
Destination	One per called station	18	TS5/ATT/PN12	486
Noise treatment	Injected noise levels set to none, low, or high	3	TS3/SPR/PN5/lo	1458
Additional delay treatment	Injected X-mission delays 0, 50, 100, 150, 300 ms	5	TS4/MCI/PN10/hi/150	7290

*Arbitrarily assigned for purposes of illustration.

where possible, ensure that the data collection is *randomized* with respect to that factor. Such randomization is generally accomplished by ensuring that there is no fixed pattern in the way that data elements are acquired. Wherever this can be done, the extra effort in test design represents a highly desirable, no-cost enhancement to the utility of the test results. In the case of the SAT, the randomization of all factors not explicitly represented in the test design can be achieved automatically, by

- Setting up a call schedule during which each test subject has designated times to place calls, perhaps assigned within blocks of time during which particular treatments are in effect.
- Determining the total set of calls that is to be made by each test subject.
- Distributing the calls to be made completed randomly within that schedule in any way that is consistent with the distribution of call attempts by treatment.

The result is a formal test schedule that looks something like Fig. 6-3, which shows a portion of an actual schedule used for a SAT. As suggested by a cursory examination of that schedule, the randomization of calling in this manner is a little more complex than it may seem from its description, because the schedule must assure that no two test subjects are calling the same destination at the same time. This means that SAT scheduling is best accomplished with the assistance of a computer program that can quickly detect and avoid conflicts. Our experience shows, however, that the rewards for the effort vis-à-vis enhanced inherent credibility and scientific defensibility of the final results are well worth the effort.

Design Effectiveness

As described earlier, the final step in test design is to create a data-collection plan that will assure, in addition to "typicality" of the sample, that test data samples will be (1) large enough to achieve high confidence levels in the estimates of the measures selected for quantification via the test, and (2) sufficient for meaningful statistical testing of differences in the measured values among different subsets of the data collected for which comparisons are *required* to answer the decision makers' questions. In doing so, for a subjective test like a SAT, however, it should also be recognized that there will be a substantial cost of acquir-

Figure 6-3 Typical service attribute test schedule for nine test subjects.

		Caller Number								
Time	Net	1:	2:	3:	4:	5:	6:	7:	8:	9:
9:30ASM	1P	HO2	AT1	AT2	MI1	MI2	HO1	NY2	DA1	DA2
9:35ASM	1P	HO1	HO2	---	---	---	NY1	MI1	---	---
9:40ASM	1P	MI2	---	---	---	HO1	HO2	---	---	---
9:45ASM	1P	---	HO1	HO2	---	---	---	SF2	AT1	---
9:50ASM	1P	DA2	HO2	AT1	AT2	MI1	MI2	NY1	NY2	DA1
9:55ASM	1P	---	---	---	HO1	HO2	---	---	---	DV2
10:00ASM	1N	HO1	HO2	DV1	DV2	SF1	SF2	NY2	DA1	DA2
10:05ASM	1N	SF1	SF2	AT1	AT2	MI1	MI2	NY1	DV1	HO1
10:10ASM	1N	MI1	MI2	NY1	DV1	DV2	SF1	HO1	HO2	AT2
10:15****	1N	****	****	****	****	****	****	****	****	****
10:20****	1N	****	****	****	****	****	****	****	****	****
10:25MAS	1N	MI1	MI2	NY1	NY2	DA1	DA2	HO1	AT1	AT2
10:30MAS	1N	DA2	HO1	AT1	AT2	MI1	MI2	NY1	NY2	DA1
10:35MAS	1N	DA2	HO1	HO2	DV1	DV2	SF1	SF2	NY2	DA1
10:40MAS	1N	AT1	AT2	MI1	MI2	NY1	NY2	DA1	DA2	HO2
10:45MAS	1N	DA1	DA2	HO2	AT1	AT2	MI1	MI2	NY1	NY2
10:50MAS	1N	SF2	NY2	DA1	DA2	HO1	HO2	DV1	DV2	SF1
10:55MAS	1N	MI1	MI2	NY1	DV1	DV2	SF1	SF2	AT1	AT2
11:00MAS	1N	DA1	DA2	HO1	HO2	DV1	DV2	SF1	SF2	NY2
11:05****	1N	****	****	****	****	****	****	****	****	****
11:10****	1N	****	****	****	****	****	****	****	****	****
11:15SMA	1N	DA1	DA2	HO1	HO2	DV1	DV2	SF1	SF2	NY2
11:20SMA	1N	NY1	DV1	DV2	SF1	SF2	AT1	AT2	MI1	MI2
11:25SMA	1N	DV1	DV2	SF1	SF2	NY2	DA1	DA2	HO1	HO2
11:30SMA	1N	---	HO1	HO2	---	---	---	DV1	DV2	---
11:35SMA	1N	SF2	AT1	AT2	MI1	MI2	NY1	HO1	HO2	SF1
11:40SMA	1N	SF2	AT1	AT2	MI1	MI2	NY1	DV1	DV2	SF1
11:45SMA	1N	DA2	HO1	HO2	DV1	DV2	SF1	SF2	NY2	DA1
11:50SMA	1N	NY2	DA1	DA2	HO2	AT1	AT2	MI1	MI2	NY1
11:55SMA	1N	DA1	DA2	HO1	HO2	DV1	DV2	SF1	SF2	NY2
12:00****	1N	****	****	****	****	****	****	****	****	****
12:05****	1N	****	****	****	****	****	****	****	****	****
12:10****	1N	****	****	****	****	****	****	****	****	****
12:15****	1N	****	****	****	****	****	****	****	****	****
12:20****	1N	****	****	****	****	****	****	****	****	****
12:25****	1N	****	****	****	****	****	****	****	****	****
12:30ASM	1N	SF2	NY2	DA1	DA2	HO1	HO2	DV1	DV2	SF1
12:35ASM	1N	NY2	DA1	DA2	HO1	HO2	DV1	DV2	SF1	SF2
12:40ASM	1N	DV1	DV2	SF1	SF2	AT1	AT2	HO2	MI2	NY1
12:45ASM	1N	SF2	NY2	DA1	DA2	HO1	HO2	DV1	DV2	SF1
12:50ASM	1N	HO1	HO2	DV1	DV2	SF1	SF2	NY2	DA1	DA2
12:55ASM	1N	AT2	MI1	MI2	NY1	DV1	DV2	SF1	SF2	AT1
1:00ASM	1N	DV2	SF1	SF2	NY2	DA1	DA2	HO1	HO2	DV1
1:05ASM	1N	HO2	NY1	DV1	DV2	SF1	SF2	AT1	AT2	MI1
1:10ASM	1N	AT2	MI1	MI2	NY1	DV1	DV2	SF1	SF2	AT1
1:15ASM	1N	SF1	SF2	NY2	DA1	DA2	HO1	HO2	DV1	DV2
1:20ASM	1N	DV2	SF1	SF2	AT1	AT2	MI1	MI2	NY1	DV1
1:25ASM	1N	HO2	DV1	DV2	SF1	SF2	NY2	DA1	DA2	HO1
1:30ASM	1N	MI1	MI2	NY1	DV1	DV2	SF1	SF2	AT1	AT2
1:35****	1N	****	****	****	****	****	****	****	****	****

Notes:

- XXD labels are abbreviations for numbers to be called
- The Net column is not shown on the caller copy; it shows which set of phones each group of 3 callers is supposed to use.
- **** denotes scheduled break
- --- indicates no call for that caller

ing each data point. Accordingly, it will be important to consciously design the data-collection plan with the two additional goals:

- Minimizing the cost of data acquisition
- Maximizing the decision utility of data acquired

These goals are in some senses competing, since the decision utility is usually enhanced by collection of more data, while the cost of data acquisition is reduced by collecting less. In the case of the SAT, the necessary balance is achieved by

- *Logical test design,* i.e., following the design process just described to decide what cells must be populated with test data in order to be able to respond to the most likely ancillary questions from persons using or challenging the test results
- *Randomization* of the other factors as just described
- *Enhancing the manipulability and interpretability of the data acquired,* by ensuring that there is a large enough sample of results in each cell to allow for analysis of effects

Except for the last one, the rationale and means of satisfying these criteria have already been addressed. In the case of enhancing manipulability and interpretability of the data, suffice it to say that the intent of the criterion is to assure that the sample sizes by category are also adequate for testing results cell by cell. This means, for example, that the procedures for developing a data-collection plan described for the case of our pollster are absolutely *not* recommended for a test like the SAT. As illustrated earlier the results from a well-designed SAT should be timeless, retaining data that can be analyzed to answer a wide variety of questions long after the proximate issue that prompted the test has been resolved. This means that we cannot follow the lead of the pollster in design of that political poll to produce a credible, scientific answer to a specific question with the smallest practicable sample size. That sample would support the best practicable estimate of the percentage of people in the country who feel one way or another about the issue. However, it would almost certainly not contain enough data in the different categories in that sample to answer questions about many regional differences, such as, "Do the people in Vermont have a substantially different opinion about the issue than the people in New Mexico?"

For this reason, we have evolved the practice of oversampling SATs, by requiring *at least 15 replications in every cell defined by the selection of factors and categories or gradations.* The rationale for this rule is that

15 represents exactly half of the nominal minimum sample size of 30 that is needed to justify the use of the normal approximation of the binomial distribution. By halving that number, we are in essence betting that we can find for any circumstance at least two categories for which data can be combined to support testing for statistically significant differences.

Examples

By following these design guidelines we have been able to satisfactorily answer nearly every question of detail in the results that has ever been posed in the 20-year history of the test while maintaining the ability to manipulate existing SAT results to make reasonable predictions of what would eventuate as a result of changes. A few of these applications will be briefly described to demonstrate the robust utility of SAT data.

Level Problems

Interestingly enough, one of the first applications of a SAT to satellite voice services revealed more about the design of the terrestrial access interface than satellite transmission. The problem was that a customer using a satellite voice service out of New York was complaining that the voice quality was "simply terrible." To respond to the complaint, we put test subjects on site at the customer location and initiated test calls into prearranged destinations in our offices in Virginia. We had little experience with the test at the time, but had already decided that what we would have to do to placate the customer was to demonstrate how the satellite service quality stacked up against the regular terrestrial telephone service at the time. Accordingly, the test was designed to include a baseline for comparison comprising regularly dialed telephone calls as well as calls completed via the satellite service.

The test resulted in impairments matrices for the satellite and terrestrial services whose comparison was very much like the hypothetical results displayed in Table 6-5, with the telling contrasts highlighted in italics. From those comparisons at least one problem was readily apparent. The installers of the satellite service failed to take into consideration the fact that the dish was on the roof of the building, so that none of the signal attenuation normally expected on the analog subscriber loops was present on the satellite accesses. The result, evinced in the impairments matrix was that the satellite signal was coming in too loud, overdriving the ear microphone.

Subsequent line testing also showed that the inbound signals via the satellite heard in New York were particularly loud, because the access circuit had been installed as an intermachine trunk rather than a central office termination requiring an insertion loss of 6 dB on the inbound side. Correction of this configuration problem greatly improved the customer's assessment of the quality of the satellite service.

Characterization of Quality without an External Basis for Comparison

Although the conscious inclusion of tests of a familiar service to create an external basis for comparison is now sine qua non for SATs, we weren't always that smart. In fact, one of the first SATs conducted nationwide to characterize the quality of satellite voice service did not include parallel testing of the terrestrial service. As a result, when we were done, we had a set of measurements for MOS for a lot of different tests run between a set of origins, $\{O_i: i = 1,\ldots,N_0\}$ and set of destinations, $\{D_j: j = 1,\ldots,M_0\}$, and no basis for testing the individual routes to determine whether the quality was significantly worse or simply about as good as we could expect.

To handle this problem we produced an internal standard by ascertaining from the available data the best MOS we could expect, under the

TABLE 6-5

Impairments Matrices for Satellite and Terrestrial Service Derived from One of the First SATs (Notional Values)

Impairment	Service	None	Some	Much
Low volume	Satellite	*1.00*		
	Terrestrial	*0.90*	*0.080*	*0.020*
Constant noise	Satellite	0.95	0.035	0.015
	Terrestrial	0.89	0.070	0.040
Speech distortion	Satellite	*0.88*	*0.070*	*0.050*
	Terrestrial	*0.99*	*0.010*	
Echo	Satellite	0.97	0.002	0.028
	Terrestrial	0.96	0.030	0.010

assumption that the $N \times M$ combinations of origin-to-destination routes sampled contained at least one that typified the best possible quality. This was accomplished by the following algorithm:

1. Calculate from the data from calls into all destinations an aggregated MOS for each origin in the original set $\{O_i: i = 1,\ldots, N_0\}$. On the basis of those results, create a subset of "good origins," $\{O_k: k = 1,\ldots,N_1\}$, by eliminating any origins for which the MOS is significantly lower than the best values.
2. For each of the destinations in the original set, $\{D_j: j = 1,\ldots,M_0\}$, calculate the aggregated MOS for each call into the reduced set of origins, $\{O_k: k = 1,\ldots,N_1\}$. Eliminate from $\{D_j\}$ any destinations for which the calculated MOS is significantly lower than the best MOS values, to create a subset of "good destinations," $\{D_l: l = 1,\ldots,M_1\}$.
3. Cycle through steps 1 and 2, eliminating origins and destinations with inferior aggregated MOS values until no more origins or destinations can be eliminated.

When this process terminates, there will remain a set of "good origins," $\{O_g\}$, and a set of "good destinations," $\{D_g\}$, for which the aggregated MOS, for instance, will represent the potential quality of the service. In addition, the set of origins remaining represents a target set that can be used to evaluate quality at particular destinations, and the set of destinations similarly represents a target set for evaluating quality at origins.

This example of handling SAT data shows that even when it is impossible for some reason to explicitly include a basis for comparison in the test design, the data collected can still be processed to develop criteria for what quality can reasonably be expected and to support route-by-route assessment to identify failures to achieve that potential.

What-if? Analysis

As illustrated by the example of the echo canceller analysis described in the earlier discussion of manipulability of SAT results, one of the most valuable features of a SAT is the association between measures of quality and the impairments that are affecting those measures. In particular, preservation of such associations in the raw data enables the examination of any number of scenarios that might be proposed. In such applications, the trick is to focus the analysis on producing distributions of "none," "some," and "much" descriptors of the basic SAT impairments as the medium for formulating hypothetical effects of treatments. This kind of focus on envisioned effects or reactions to impairments rather than direct estimates of quality make it a lot easier to both formulate and test the sensitivities of hypotheses. For example, suppose that a manufacturer is proposing a device that is claimed to eliminate 90 percent of the noise on telephone calls. The problem of assessing the value of such a device might otherwise be daunting but can be readily attacked by manipulation of SAT data by

1. Retrieving historical SAT data collected from the environment in which the device might be deployed
2. Determining the distributions of impairments in that data, to produce profiles like those shown in Table 6-5, together with the MOS and P[UDI]
3. Describing reasonable expectations of the effect of the device on the expected incidence of reports of constant noise
4. Extracting a sample from all the SAT data whose incidence of reports of constant noise matches the hypothetical values
5. Calculating the MOS and P[UDI] from the sample constructed in step 4 to produce an estimate of the expected effects of the noise reduction device on quality and usability of the service.

Such an analysis process produces results that have a lot of inherent credibility, because (1) the assumptions of the effect are a lot more intuitive than guesses of direct effects on opinion scores; (2) values estimated in step 5 are derived from known actual responses from test subjects, rather than extrapolations from models that "estimate" the responses; and (3) whatever results are produced can be readily tested for stability by examining the estimates produced by any number of different assumptions in step 3.

SAT-Based Analysis Tools

To enhance the utility of SAT data, development of the SAT technology over the last 2 decades has included production of a number of analysis tools to facilitate what-if analyses of the kind just described. These include:

1. *Characterization of toll quality.* To compare its service quality with the competition, MCI used to run an annual nationwide SAT. The last comparative SAT of this kind produced data on more than 26,000 calls, distributed over 254 different origin-to-destination routes and three different networks, evaluated by 116 different callers. The long-distance system characterized in that test was all-digital, utilizing 64-kbit/s PCM-encoding to transmit voice signals over optical long-distance transport protected by active echo cancellation. The results of that test, which showed essentially no difference among the DDD services of any of the three major carriers, can reasonably be surmised to fairly typify the best voice quality that can be expected by users of U.S. domestic long-distance telephone service. Values of measures derived from that test can therefore be used as the benchmark against which the likely user perception of voice quality can be gauged.

2. *A basis for predicting SAT scores for untested environments.* In order to make use of the extensive historical data from SATs for purposes of analyzing the likely impact of changes in incidence or severity of impairments, data from the 5 years of SATs were processed to produce the TSAT model (referred to at the end of the section on Measures and Comparisons). This model is based on a sample in excess of 150,000 evaluated calls, which was analyzed to determine how user assessments of quality and effect varied with their assessments of the severity of low volume, noise, speech distortion, and echo. The resultant model supports credible, internally consistent estimation of the likely measures of quality that would result from a SAT in which any particular distribution of incidence and severity impairments were experienced. In tests of this application, for example, actual and TSAT-predicted mean opinion scores typically differ by less than 0.1.

3. *Heuristics for interpreting differences in SAT quality measures.* Effective use of the results of SATs over the last 2 decades has validated heuristic rules of interpretation of SAT scores to identify conditions under which users will perceive a noticeable difference in the quality of voice between two services, or can be expected to complain that voice quality has become unacceptably degraded. These rules were evolved empirically, by observing conditions and/or situations in which users were complaining that there was a noticeable difference between older service and newly installed SBS/MCI service or that the service characteristics were unacceptable, and then stopped complaining after changes were made. They provide a valuable adjunct to SATs in the form of interpretation aids that can be used to translate test results into information.

Like many other technical details that will be alluded to in this book, the specifics of these analysis tools are proprietary or confidential, because they are key elements in commercially licensed software and hardware. More general descriptions like these, which do not reveal such elements, are included to convey a sense of what can and has been done to make the techniques described here reliable and useful.

CHAPTER 7

Measurement of Speech Distortion

As described in Chap. 6, subjective user tests have been around for a long time, yet they can still serve us well as a means of accurately gauging likely user perception of voice quality. However, because of the necessary involvement of a relatively large number of test participants and the extensive sampling requirements, such tests tend to be labor-intensive and relatively expensive. Consequently, one of the major goals of developing test and evaluation methodology for voice quality has been to accomplish the same ends by means of analysis of data that can be acquired quickly and inexpensively via objective measurements.

In this chapter we examine some of the earliest attempts at doing this. These efforts focused on ways of measuring how much speech waveforms are distorted during the course of their injection, encoding, transmission, decoding, and extraction from a telephone system. The objective was to find a measure that would exhibit a high correlation with user assessment of quality, thereby supporting accurate prediction of how users would gauge the quality of telephonic speech.

A *Very* Short Course in the Physiology of Speech and Hearing

The focus of such efforts to measure and analyze speech distortion was on the speech waveforms delivered to the distant speaker. Since these waveforms represent what was heard, the user-perception models of this kind were usually based on studies of the generation and cognition of human speech. It is therefore appropriate to begin here by recounting some of the concepts involved. This will take us into areas of human anatomy and physiology and the physics of sound. However, don't panic! What we will be going through here is a *very* short course.

Speech Production

As all those who are listening to the words in their mind as they read this book can tell you by actual test, intelligible human speech is generated by particular patterns of sound. Those patterns comprise elements that can be individually formed and combined to create the syllables that can be combined to create the words that convey thoughts or evoke mental images. The basic building blocks in this hierarchy are called *phonemes.* These represent vocally articulated sounds that exhibit enough commonality to be perceived to be the same, regardless of the possible variations in frequency and volume of the voice of the speaker. The International Phonetic Alphabet as revised in 1993 identifies something like 107 basic phonemes comprising 68 consonant sounds, 28 vowel sounds, and 11 others. The exact numbers and sounds here are not important. What is important, however, is that there are very few basic phonemes, relative to the myriad sounds that might be articulated by the human voice.

We learn to recognize these particular building blocks by hearing them in words and learning how to form them in our own speech. We usually begin this process by learning how to form the relatively small number of phonemes used in the language of our parents. We later may go on to add the ability to recognize and articulate other phonemes in the course of the arduous effort of learning how to speak a foreign language or a regional dialect that contains phonemes that we're not used to forming.

Articulation of the particular sounds that are necessary to form syllables and words in any language begins with the human vocal cords. These are a pair of membranes about an inch long stretched across the nearly rigid frame of the larynx (commonly recognized externally as a person's Adam's apple). When the membranes are not tensed, breath passes through them without impediment. When we want to talk, we tense the vocal cords, creating a flexible cover across the larynx with a slit in it. The cover is sometimes referred to as the *glottis,* while other times glottis is used to refer to the slit.

Either way, the glottis created by tensing the vocal cords creates an impediment to the flow of air being exhaled by the lungs, causing air pressure underneath the cover to increase, while reducing the air pressure on the side above the cover. The air pressure underneath the cover then builds up to the point that it is strong enough to push through the cover, thereby creating a burst of air pressure above the cover. This buildup and release repeats many times a second, creating a sound wave above the cover, behind and in the mouth, where it can be let out when we open our mouths to speak (Fig. 7-1) or held in for humming, 'om'-ing and 'umm'-ing while the mouth is closed. The buildup and release of air pressure by the vocal cords thus produces the same kind of variations in air pressure caused by vibrations in a rigid object that alternately compress air as they push it forward and decompress it as they move back. Whenever an up-and-down or back-and-forth vibration of this kind occurs, one segment of the variations in which the change

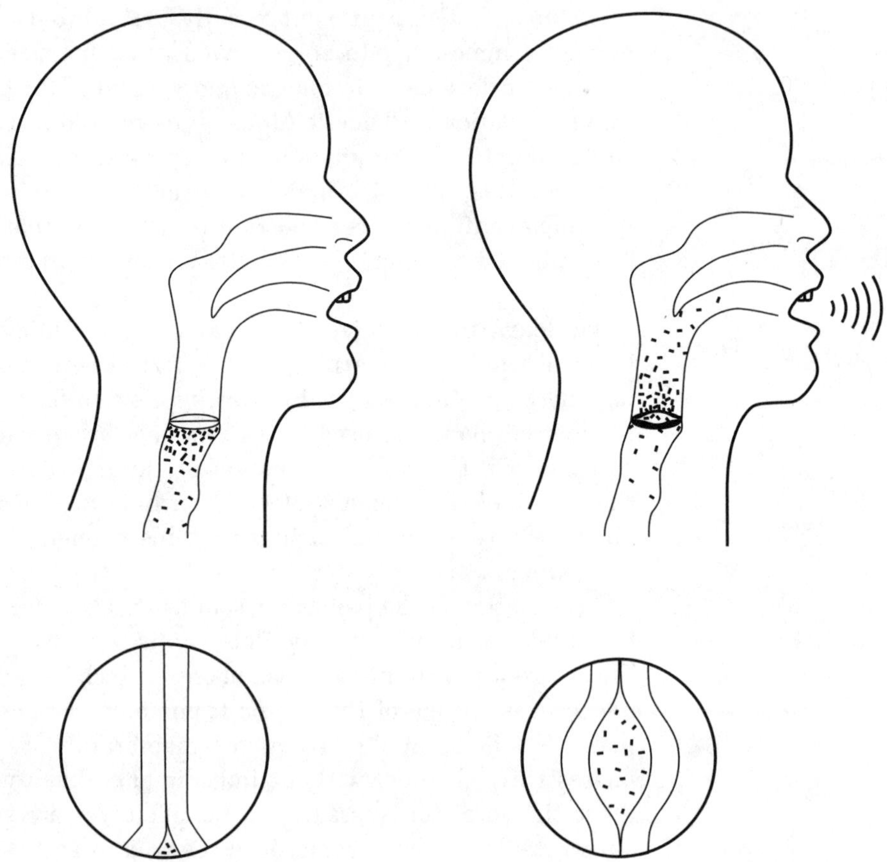

Figure 7-1 Production of fundamental frequency by air flow across the vocal cords.

goes from a maximum point to a minimum point and back to a maximum is called a *cycle.* When the variation is a regular vibration, causing the maximum and minimum points to occur at regular intervals, the intervals are usually measured in number of cycles per second, and expressed in hertz (Hz), where 1 Hz = 1 cycle/s.

The number of cycles per second in a regularly varying vibration is referred to as its *frequency.* The frequency of the sound wave created by the vocal cords at any point in time is called the *fundamental* frequency. Since this is a very short course, we will simply observe here that this fundamental usually ranges between 80 and 200 Hz in males and between 150 and 300 Hz in females. Changes in the tension on the vocal cords change this frequency, with greater tension producing higher frequencies. (Notice that the adherence to the very short course format here allows us to avoid the question of "why?".)

In addition to setting up a fundamental frequency with the vocal cords, we can form phonemes by allowing our breath to flow unimpeded across the vocal cords and shaping sounds in other ways. For example, we can create an *s* sound by blowing air out of our lungs and setting up vibrations across a stricture in the mouth between the tongue and the back of the front teeth. We form a *q* sound similarly by setting up vibrations in a stricture between the tongue and the roof of the mouth. In the case of the initial *q* sound, any change in the shape of the lips creates a following vowel sound without the involvement of the vocal cords. Rounding the lips forward produces the inseparable English *qu,* creating what is thought to be a spelling rule, but is, in fact, a phonetic artifact. Stretching the lips turns the same initial sound wave into something that transitions to a hard *k,* explaining why Spanish speakers greet each other with "k-passa" instead of "q-passa" when pronouncing "*¿Que pasa?*" We form the *p* sound by holding air flowing freely from the lungs back behind closed lips and releasing it to generate a "pulse" of air compression.

To articulate a phoneme, then, we

- Tense the vocal cords to initiate a sound wave with a particular fundamental frequency, or relax them to allow an unimpeded flow of air and create other strictures to build up and release air pressure
- Shape the resultant sound wave by the manner in which we hold the lips, tongue, and jaw
- Vary the volume of the wave by controlling the amount of air coming out of the lungs

This process produces what we will call the *basal waveform* for a phoneme, comprising one cycle of a waveform that will be repeated many times, perhaps with variations in amplitude, to form a phoneme.

We learn how to form the basal waveforms to articulate particular phonemes by experimenting with positioning of the lips, tongue, and jaw to shape the oral and pharyngeal cavities, practicing until what we hear from ourselves sounds like what we hear when other people are speaking. (Of course, as anyone who has suffered through listening to themselves on a recording will attest, what we hear when we speak is not at all like what those outside of ourselves hear. But it does get to be close enough so that those who learn to speak in Boston, New York, Iowa, and London eventually make a sound for the *a* in "father" that is enough like everyone else's to be recognized as a short-*a* sound, or, at least, something clearly *intended* to be that.)

To articulate syllables we then string phonemes together in a sequence, *dwelling* on each long enough to allow for the particularly formed sound representing an individual phoneme to set up a distinct pattern of repetition of basal waveforms. As illustrated in Fig. 7-2, this creates a pattern in the articulation of a syllable in which

1. Articulation begins with an *attack* on the basal waveform for the first phoneme from a point of silence or a drop in volume from the ending of the previous syllable.
2. Each phoneme is formed by *dwell* on the appropriately shaped sound, during which time the amplitude of that sound may gradually rise up to a holding level and then slowly taper off as the articulation ceases.
3. There are *transitions* in sound and shape of waveforms or very short silences between successive phonemes, until syllable completion occurs.
4. Syllable *completion* occurs with a trailing off of the sound energy in the last phoneme into the attack on the next one.

In this model of the production of syllables, the dwell time on each phoneme is on the order of 10 to 120 ms (1 ms = 1/1000th of a second). If the dwell is held for longer than normal, the phoneme begins to sound musical. Transitions are on the order of 1 to 10 ms and may involve modulation to a different fundamental frequency. When it is possible to shape the mouth to go from one phoneme to another continuously, without a noticeable transition, the two consecutive phonemes sound like one longer phoneme that is described by linguists as a *diphthong*. The attack on a leading phoneme in the initial syllable of a word is manifested as a gradual rise in volume over 10 to 20 ms as the vocal cords begin to vibrate, or as a sudden, discontinuous burst of sound when the initial phoneme is a *plosive* (e.g., *b*, hard *c*, *d*) or is created by blowing air out of the mouth with a sharp push, without vibrating the vocal cords (*p* sounds; or *h* sounds at the beginning of words, except in Spanish and Cockney, in which initial *h*'s are just dropped).

Syllabic completion is marked by a decrease in sound energy, as we mentally tool up the mechanism to string together the phonemes that make up the next syllable. These intersyllabic dips in volume and transition to the attack on the next syllable take on the order of 5 to 25 ms. The pattern is the same at the ends of words, but somewhat longer. In the case of rapid speech, however, there is so little difference between the completions of syllables and words that persons unfamiliar with the

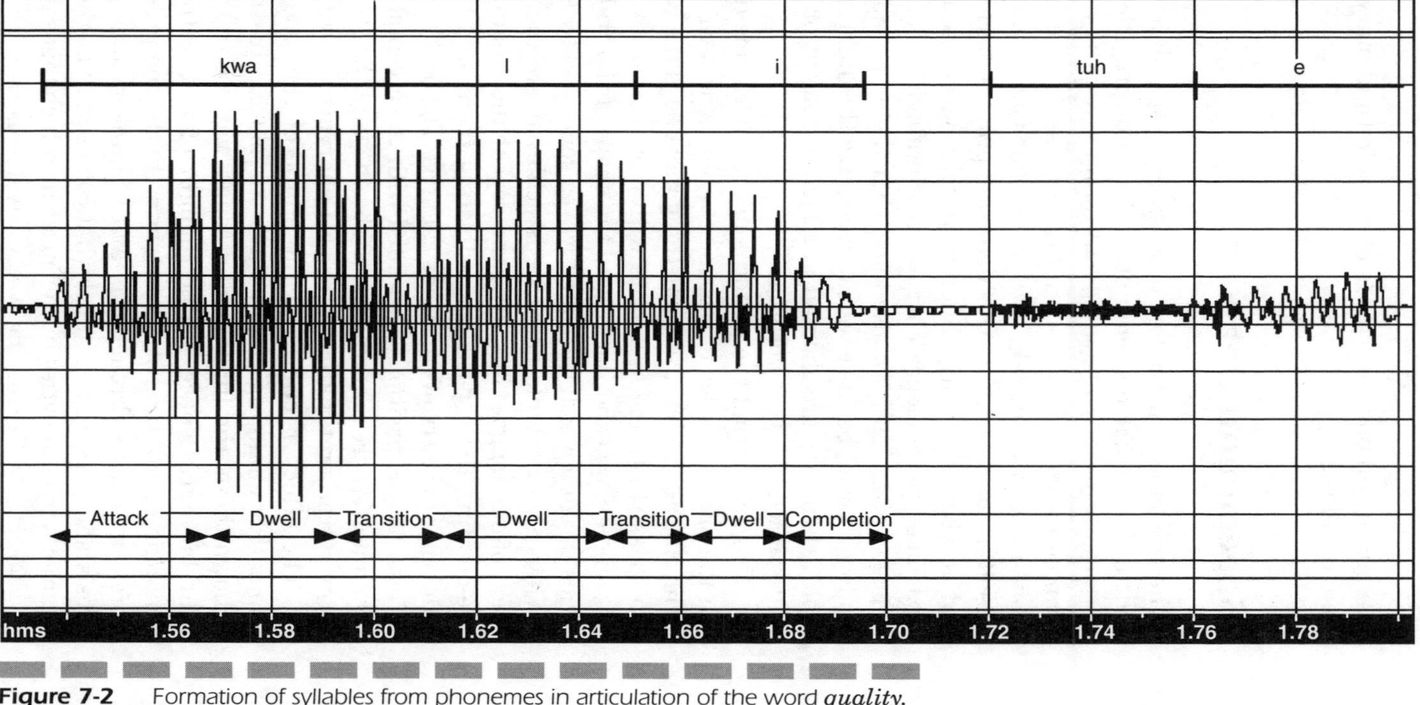

Figure 7-2 Formation of syllables from phonemes in articulation of the word *quality*.

language will only hear ends of syllables and marvel at how fast everyone is talking.

Speech and Hearing

The point to the preceding description of speech is that what is presented to the ear for discrimination and interpretation when a syllable is uttered is a series of articulated sound waves that exhibit patterns that can be recognized by the brain to represent, for example, an articulated "oh" instead of an "ah." We know that the fundamental frequency by itself is not material to this detection process, because an "ah" articulated at what are clearly different fundamental frequencies, including the extremes of frequency encountered in singing, is still heard as an "ah" instead of an "oh." We also know that the amplitude of the sound waves is immaterial, because we can recognize the difference between "ah" and "oh" when they are shouted in a loud argument or spoken very softly in utterances of "sweet nothings."

This means that the patterns in the waveforms that are discriminated in the brain in order to recognize a phoneme have to be something else. The clue as to what that something must be is the way that we learn how to speak. In trying to imitate with our own voices what we are hearing around us, we experiment with the shape of the oral and pharyngeal cavities through which the sound waves generated by the vocal cords travel. When an acceptable phoneme is achieved, we then remember the position of our tongue, lips, etc., and repeat the creation of the phoneme by re-creating those positions, adjusting them slightly as we grow larger and the geometry of our heads changes. The only thing that is common to those various positions associated with the acceptable phonemes is, then, the interference patterns caused by the way that the fundamental sound wave reverberates around the oral-pharyngeal cavity.

The effect of changing the shape of the cavity through which a smoothly changing sound wave travels is illustrated in Fig. 7-3. As shown there, the fundamental sound wave set up by the vocal cords in the laryngeal space can be thought of as creating a fairly smoothly changing amplitude of air pressure. As the compression front of that air pressure passes through the cavity, however, some parts encounter more resistance, others less, until what comes out has the same frequency but exhibits irregularities in the once smooth changes in amplitude.

Now, the difference between waveforms like the smooth one shown in Fig. 7-3 and the one with irregular amplitudes is the *frequency spectrum*

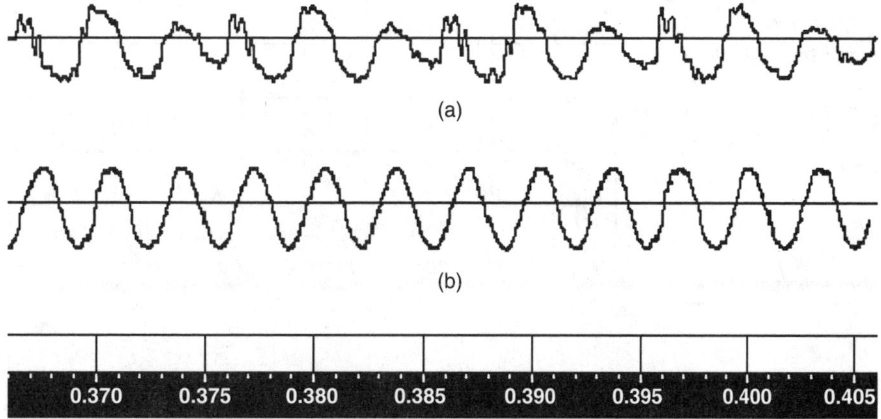

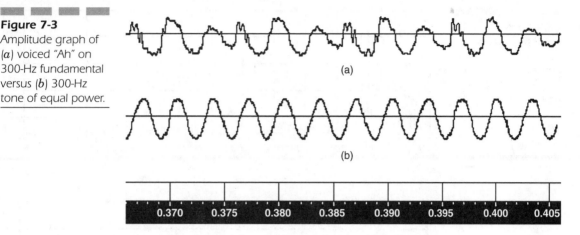

Figure 7-3 Amplitude graph of (*a*) voiced "Ah" on 300-Hz fundamental versus (*b*) 300-Hz tone of equal power.

of the waveform, which is a representation showing how much of the total energy is associated with different frequencies. The idea of such a spectrum comes from mathematical theorems proved a long, long time ago by a mathematician by the name of Fourier. He showed that any mathematical function representing a repeating waveform can be described as the sum of smoothly changing waveforms of basic frequencies (represented by sine and cosine functions) that are each multiplied by some weighting factor representing the proportion of the energy that the waveform contributes to the whole. Thus, because the lower waveform in Fig. 7-3 has but one frequency component, it can be expressed as having a weighting factor of 1 for the sine function at 300 Hz, and all other components wind up with a weighting factor of 0. The waveform with the same frequency, but irregular amplitude, shown above it is the result of a sum of a whole lot of different sine and cosine functions for different frequencies, each with a different weighting factor. The profile of weighting factors for each frequency that contributes energy to its whole is called the frequency spectrum of the waveform.

Such spectrums for the waveforms in Fig. 7-3 are displayed graphically in Fig. 7-4. These were created by a mathematical process, called a *Fourier transform,* whereby the amplitude of a waveform at any point in time is equated to the sum of products of weighting factors applied to sine and cosine functions with frequencies 1, 2, 3,…,N Hz. Enough equations are then written to enable solution of the equations for the coefficients, which then become the weighting factors for the N discrete frequencies used in the representation of the spectrum. For the spectrums shown in Fig. 7-4, N was 4000, so these transforms involved solution of

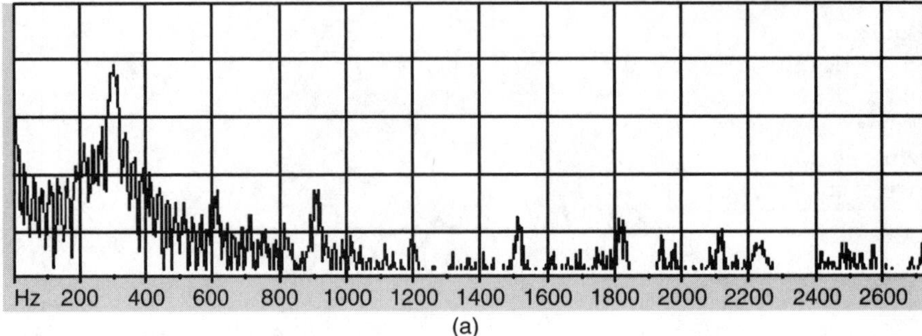

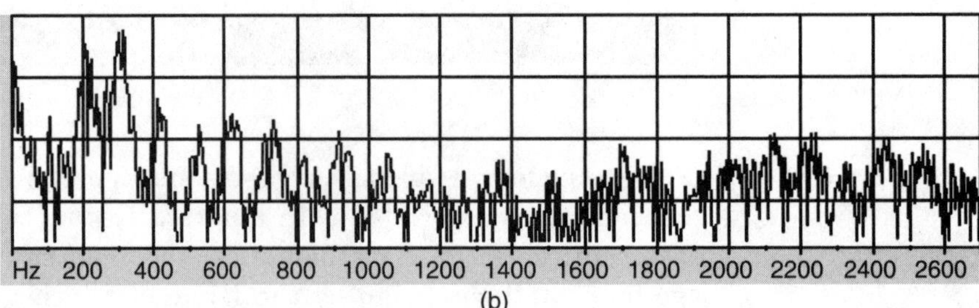

Figure 7-4 Effect of vocalization on 300-Hz tone. *(a)* Spectrum of 300-Hz tone digitized with 64-K μ-law PCM. *(b)* Spectrum of voice signal with 300-Hz fundamental with same digitization.

8000 simultaneous equations for 8000 unknowns. Even so, what we got is but an approximation to the real transformation succinctly defined by Fourier with fewer than 20 symbols to involve an infinite number of coefficients. The errors due to our use of finite approximation for an infinite series accounts for the fact that the spectrum for the sine wave shows a sharp peak at 300 Hz, rather than a point spike. Similar errors produce the "grassy" floor of both spectrums, indicating the presence of some energy in each of the frequencies.

Hearing

The mechanics of speech production thus suggest that the spectrum of the sound waves produced when a phoneme is articulated is the distinctive characteristic of phonemes. Such sound waves will have a nearly constant fundamental frequency but will have irregular amplitudes caused by the shaping. Moreover, because of all the possibilities for shaping, we would expect that these sound waves would have to be repeated with approximately the same shape for some number of repetitions before they could be distinguished. Figure 7-5 lends credence to this conclusion by display-

Figure 7-5 Articulation of the consonant *n* by different speakers.

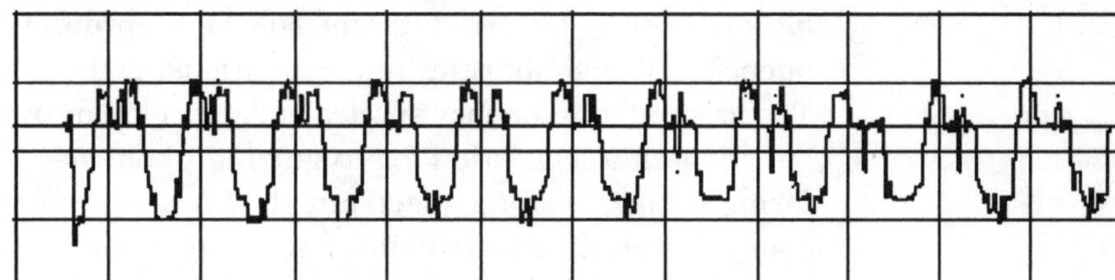

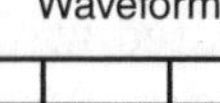

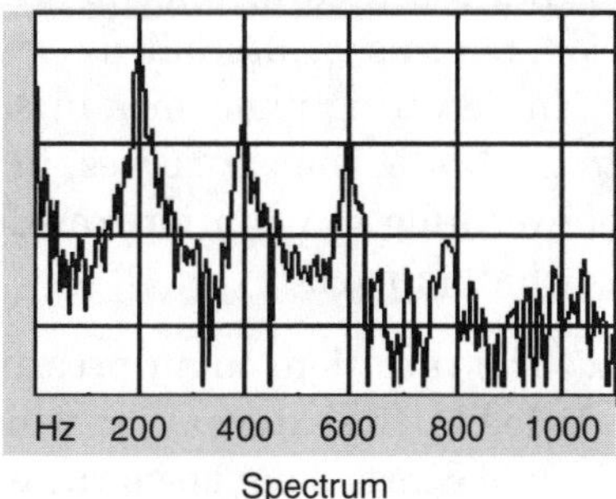

SPEAKER B

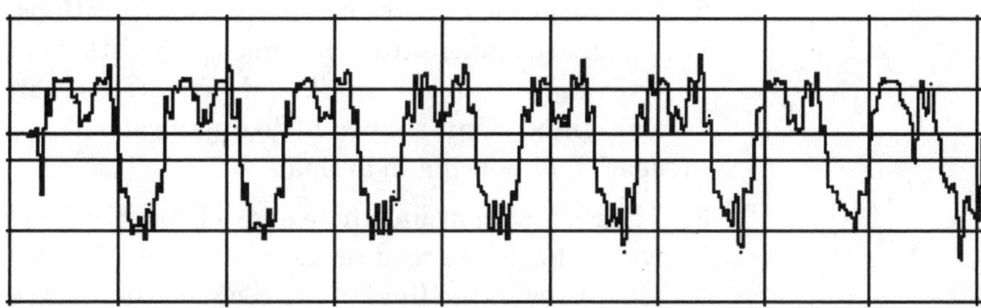

Waveform

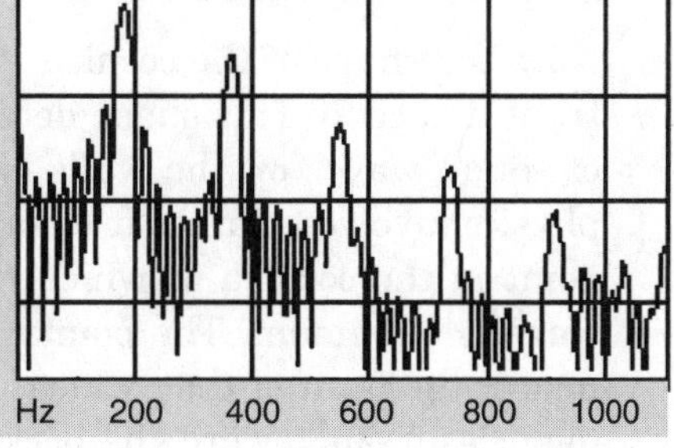

Spectrum

ing a repetition of the consonant *n* in the first syllable of the word *information* spoken by two different speakers on different fundamental frequencies. The similarities and repetitions in these shapes are apparent. The associated frequency spectra suggest, moreover, that the likely distinguishing characteristic for recognizing phonemes is a pattern of waveforms with the same spectrum that are strung together over a long enough dwell time to exhibit a number of repetitions. Phonemes are then discriminated by the arrangement of the peaks in energy in the spectrum of that repeated waveform. (There is a lot more evidence for this conclusion, but this is, after all, a *very* short course.)

In detecting phonemes in speech and communicating their presence to the brain, the ear serves, in essence, as a biological computer capable of executing the Fourier transform illustrated in Figs. 7-4 and 7-5. Here is what happens:

1. The variations in air pressure representing a sound wave are amplified by the external ear which acts as a funnel to collect sound waves and direct them down the ear canal to a receiving membrane called the ear drum. (The ear drum is more formally known as the *tympanic* membrane.)
2. The variations in air pressure set up vibrations of the ear drum, which are transmitted via some fancy little bones (with the suggestive names *hammer, anvil,* and *stirrup*) to create vibrations in the fluid-filled, snail-shaped little organ called the *cochlea* that is the seat of all hearing in animals.
3. The cochlea contains thousands of "hair" cells that stand along its walls like submerged reeds in water. As the sound waves cause vibrations in the fluid of the cochlea, the hair cells get pushed together with compression and pulled apart by the following decompression.

Now, the shape of the cochlea is a spiraled chamber that unwinds into a truncated cone. This shape determines the effect of different frequencies of sound waves on the walls of the cochlea. Without going into all the physics involved, suffice it to say that changes in frequency will shift the point in the cochlea at which the amount of compression from a sound wave is maximum. The points associated with higher frequencies occur nearer the front of the cochlea, and the points associated with lower frequencies occur toward the back. Each little group of hair cells thus has an associated frequency that pushes them together the most. In a normal human ear there are hair cell groups that can be differentially excited by frequencies ranging from 20 to 20,000 Hz. Within this range, all persons with normal hearing are expected to be able to discriminate tones between 250 and 8000 Hz at approximately the same power level. On either side of this range there is a gradual decline in hearing sensitivity, so that we may be able to hear very low frequencies or very high ones in the 20- to 20,000-Hz range, but they have to be very loud to be audible.

When a group of hair cells is pushed together, there is a bioelectric impulse generated that changes the electrical potential in attached nerve endings proportionally to how much they are pushed. Enough of a change in that potential then triggers a nerve discharge that runs along the auditory nerve group to tell the brain that a particular frequency component has been detected in the sounds being heard. This means that if a sound wave retains its basic spectrum over a long enough time period, then there will be a distinct difference in the total amount of excitation of the hair cells caused by different frequency components. The accumulated differences then produce a biological equivalent of the spectra as shown in Fig. 7-6, in which spectrum B in Fig. 7-4 has been annotated to show what each group of hair cells would be communicating.

The ear thus supports a phoneme recognition process based on the repeated excitation of the same groups of hair cells over time to capture and relay the content of the frequency spectrum of sound waves hitting the ear. As illustrated in Fig. 7-2, which shows attacks and phoneme-to-phoneme transitions in the pronunciation of the word *quality,* this detection process is probably facilitated by naturally occurring changes in volume. The dips in volume that occur as the articulation shifts from one

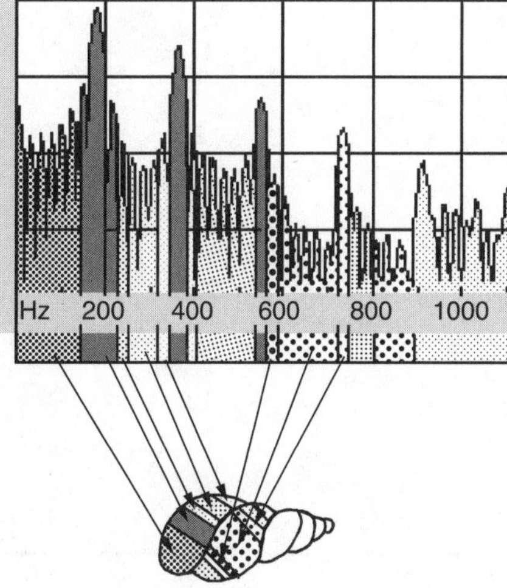

Figure 7-6
Excitation of hair cells by speaker B's articulation of *n.*

phoneme to another and during deliberate pauses between words may provide subtle cues to listeners that something new is on the way.

Implications

The inclusion of this *very* short course in the physiology of speech and hearing is a quick and convenient way of introducing many of the facts and notions on which models developed to measure perceptible speech distortion are based. Although we have provided only a simplified description of a complex process, it does facilitate understanding of dynamics of speech illustrated in Table 7-1. Appreciation of the expected timing and duration of periods of attack, articulation of phonemes, transitions between phonemes, and syllable completion shown there are critical to the understanding of how various waveform encoding and transmission schemes may affect user perception of voice quality.

The brief overview of what is going on in hearing also begins to suggest notions of what might constitute a distortion of a speech waveform.

TABLE 7-1

Timing of the Articulation of the Word *Information*

		Timing, ms			
Syllable	**Phoneme**	**Attack**	**Dwell**	**Transition**	**Completion**
1	I	16	25	8	
1	N		44		
					25
2	F	23	37	7	
2	O		11	1	
2	UR		20		
					6
3	M		64	5	
3	A	24	63	23	
					10
4	SH	16	67	12	
4	I	5	30	5	
4	N		63		15

If the perception of phonemes is based on the spectrum of the sound waves generated at the earpiece of a telephone, then we must be concerned with effects that change the waveform spectra introduced at the distant mouthpiece. The principal possible effects are as follows.

Filtering Studies have shown that the environmental noise captured by the mouthpiece on a telephone handset tends to be in the low-frequency ranges. To avoid the transmission of such noise, the electrical signals from the mouthpiece are by design filtered to cut out components of the frequency spectra below 300 Hz and above 3400 Hz.

Masking As a telephone signal is transmitted, various processes can add spurious energy to the electrical signals that are intended to emulate the voice waveforms. For example, random analog line noise will add energy across the spectrum that will create the grassy characteristics spectrum noted in Fig. 7-4. If those additions to the spectrum are great enough, they begin to interfere with the detection of the peaks in the frequency spectrum that represent the semantic content of speech waveforms.

Aliasing The processing of the sound waveforms introduces spikes in the spectrum that represent the presence of significant amounts of energy in frequencies not in the speech waveforms. For example, the so-called quantizing noise, representing the error in sampling and digital encoding of the amplitude signals through a PCM codec can introduce whole patterns of such spikes across the spectrum. If they are not masked by random noise, these spikes will not be great enough to interfere with the ear's phoneme detection process. However, they will change the perceived quality of the voice, making the distant speaker's voice sound raspy or like a speaker is talking through a buzz. Other sources of aliasing in a telephone system include

- *Overamplification* of analog signals, which will produce so-called *intermodulation* effects that are, in fact, aliasing of frequencies in the spectrum of the signal.
- *High-power levels* into PCM codecs, which may produce frequencies in the spectrum of the speech waveform whose patterns are uniquely determined by the components in the spectrum that have the highest power (see, e.g., Fig. 7-7), which are, unfortunately exactly the ones the ear keys on.

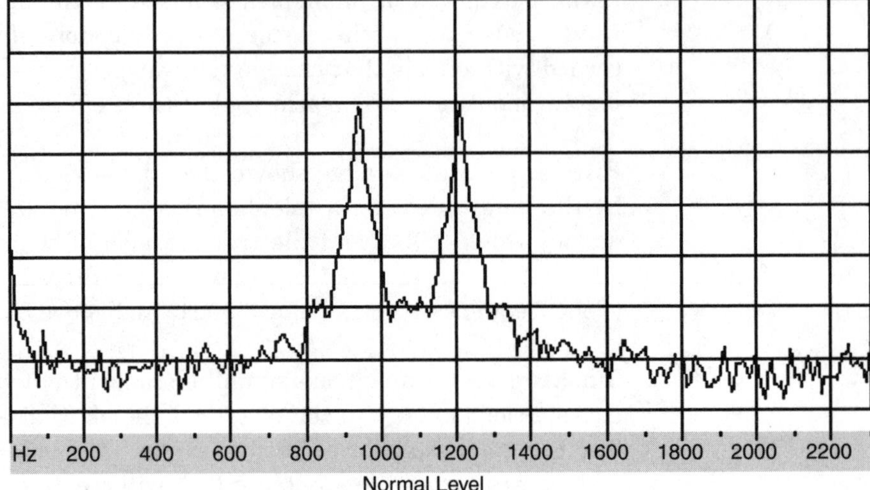

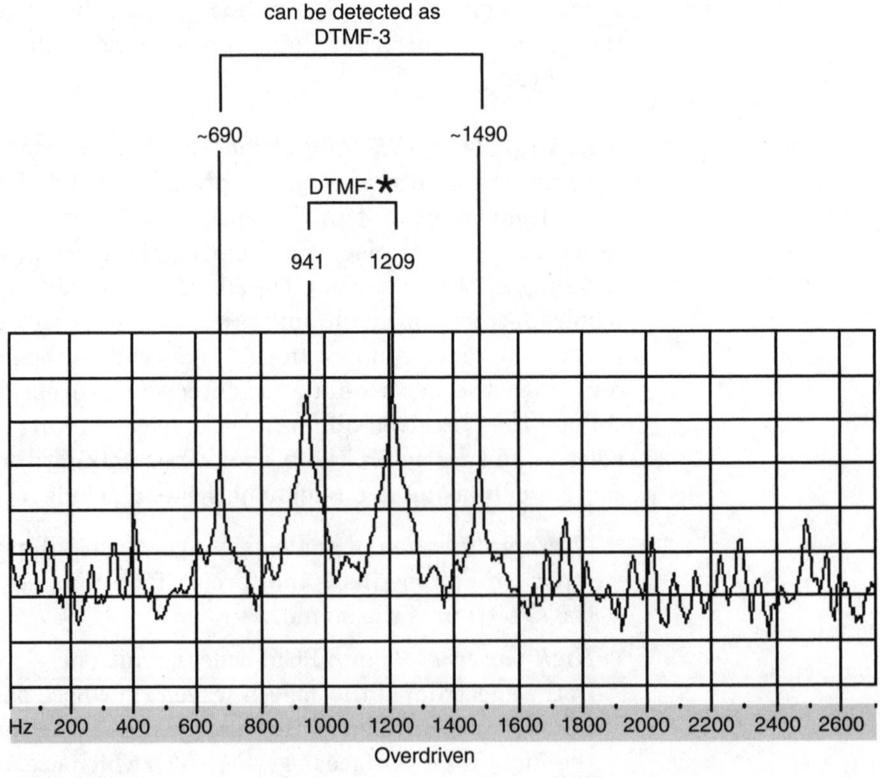

Figure 7-7 Effects of overdrive on reception of DTMF-*.

- *Interference* comprising addition of signals that have a very distinctive frequency spectrum, like the 60-cycle "hum" that was of such concern in the early days of telephony (way back in the 1950s).

Waveform Distortion from Digital Signal Errors Finally, when speech waveforms are digitized and carried over digital links, any error in the digital data transmitted is going to produce a change in the shape of the speech waveforms. Such changes can have profound effects on the frequency spectra that the ear must produce in order to support detection and recognition of phonemes.

Other possible sources of distortion of speech suggested by our descriptions of the processes of speech and hearing include the following.

Sterilization As illustrated in the speech waveforms in Fig. 7-5 and described in our models, natural speech includes smooth changes in the volume of articulated phonemes, in the attack on the initial phoneme in a syllable, in some of the transitions between successive phonemes, and in the completion of syllables and words. Such smooth volumetric changes do not affect the spectrum of the waveforms for the phonemes, but they do create the "smooth" quality that makes speech pleasant to listen to. When those changes are taken out of natural speech, the result is a robotic-sounding language—a harsh, mechanical, monotone speech that is totally lacking in warmth or expression.

Clipping As described in Chap. 4, one of the phenomena associated with modern digital telephony is the clipping of the attack in initial phonemes and completion of syllables that can occur when the transmission system is configured to use voice activity compression. When such techniques are applied, the only signals transmitted are those whose signals are above a given threshold of power. Depending on the design of the voice activity detection routine, extreme cases of this kind of clipping can also affect intrasyllabic speech waveforms.

Omission Another phenomenon, uniquely associated with packet switching, is the possibility of dropping packets, which can result in gaps in the sequence of speech waveforms. The gaps may be filled artificially, or simply appear as nulls, without changing the frequency spectrum of the segment affected. These omissions at the distant codec can nonetheless produce artifacts in the form of "pops" and warbling speech.

Types of Speech Distortion Measurement

The implications from studies of speech and hearing described here were recognized far earlier than they could actually be exploited for measurement and evaluation of voice quality. As late as 1980, the best attempts to measure some of the more complex types of waveform distortion that might affect user perception of voice quality were based on mechanical simulators like the artificial ear for telephonometric measurements patented in 1978 (see App. D). More robust measurement capabilities had to await the evolution of microcomputing technology that could handle the massive computational workloads involved in capturing, storing, and processing speech waveforms to quantify complex measures of waveform distortion.

As those capabilities have become available, however, the development of models that would predict user perception of speech quality based on analysis of deformations of speech waveforms has become a distinct possibility. At the core of these efforts have been attempts to define measures of speech distortion that

1. Can be quantified by processing of digital images of complex waveforms
2. Reflect differences between natural and telephonic speech that affect what users hear without rendering speech unintelligible
3. Can be shown in subjective user tests to exhibit a high correlation between values of the measures and user assessments of voice quality

Various efforts in this direction have resulted in four different kinds of measurement techniques, depending on whether the measurements involved are *psychoacoustic* or *electroacoustic* and on whether the data acquisition is *active* or *passive*. These distinctions are described in Table 7-2, together with the names of the particular techniques that are used here to illustrate each approach.

In the terms defined in Table 7-2, our description of the techniques begins with the active test versions and ends with the description of the passive electroacoustic technique. This order of the presentation approx-

TABLE 7-2

Types of Speech Distortion Measurement

	Active (based on comparison of what was received with what was transmitted)	**Passive** (based only on what is received, without knowledge of what was transmitted)
Electroacoustic (based only on the characteristics of the electrical waveforms)	Multifrequency tone tests	Analysis of statistics of rates of change in PCM signals
Psychoacoustic (based on transforms of acoustic waveforms that model human hearing)	Perceptual Speech Quality Measure (PSQM)	Extensions of PAMS
	Perceptual Analysis Measurement System (PAMS)	
	Perceptual Evaluation of Speech Quality (PESQ)	

imates the order of development reflected in dates of the associated patents listed in App. D. As elsewhere in this book, the descriptions here are intended to explain how the models are constructed and discuss the implied strengths and weaknesses for particular applications to measurement and evaluation of the quality of packet-switched voice services and not to provide information on how to implement them. More technical detail from the public domain can be found in the patent extracts in App. D, in references cited in those patents, and in other locations cited in this book. However, parameters, equations, and specific algorithms that would enable someone to construct and use any of these models are the intellectual property of their developers and are therefore not appropriate subject matter.

Active Measurement Techniques

Active measurement techniques are based on transmission of waveforms with known characteristics, capture of the received image of that waveform at the distant end of a telephonic connection, and processing to compare the two images and quantify measures defined in terms of differences between what is transmitted and what is received. Their distinguishing quality is the fact that the test waveform transmitted can be created for purposes of testing. This makes it possible to base the measurement on expected differences between the original waveform that has been designed to almost any specification. Such test waveforms can, for example, be tailored to typify almost any language or to enhance the ability to detect and measure distortions of particular interest.

Electroacoustic Experiments

The earliest attempts to capitalize on the idea of analyzing the received image of a tailored waveform to produce measurements that might serve as indicators of likely user perception of voice quality were electroacoustic, based on capabilities to generate waveforms whose electrical image was readily measurable and well understood. Two examples of this genera follow.

21- and 23-Tone Multifrequency Tests One of the earliest applications of the notion of testing based on transmission of waveforms designed to support measurement of specific kinds of deformations is the 21-tone test developed by AT&T. This particular test was produced as an embodiment of the more general Adaptive Multi-Tone Transmission Parameter Test Arrangement patented in 1981, and later distributed as a revised standard 23-tone test. These multifrequency test signals were actually designed for the language of modems, principally to support measurement of the intermodulation distortion (creation of energy in a third frequency when two frequencies are transmitted at nearly the same time) that can cause major problems in transmitting signals utilizing frequency- and phase-shift keying. The signal transmitted was a complex waveform generated by specifying a set of frequencies equally spaced across the spectrum from about 150 to 3300 Hz and assigning equal weighting factors to each, thereby producing a sample of the frequency response across the nominal voice channel spectrum.

Because of its structure, the spectrum of the received version of such signals is expected to exhibit peaks at the frequencies transmitted like those shown in Fig. 7-8, showing the shape of the transmission filter. Any significant amount of energy outside of that pattern indicates excessive waveform deformations due to such factors as

- Envelope delay (a phenomenon that changes the relative phase of different frequencies in an electrical signal, because they do not arrive at the same time, even when they are transmitted at the same time)
- Unequal changes in the power of the different frequencies, causing differences in the weighting factors for the different frequencies in the spectrum of the multitone signal
- Analog transmission characteristics that cause jitter (i.e., rapid variations) in the frequency, phase, or amplitude of transmitted signals
- Significant quantizing noise from digitization of the signal

The principal measure calculated from the received image of the multifrequency test signal is the signal (power)-to-distortion (power) ratio, expressed in decibels. The power measurements in this case make use of the *inverse Fourier transform,* which takes a frequency spectrum and produces the description of the waveform. The signal power is calculated by applying an inverse Fourier transform to only those portions of the received signal spectrum that are close to the frequencies in the transmitted tone. The measure of the distortion is the measure of the power in the signal obtained by applying an inverse Fourier transform to the rest of the spectrum and calculating the power in the time domain.

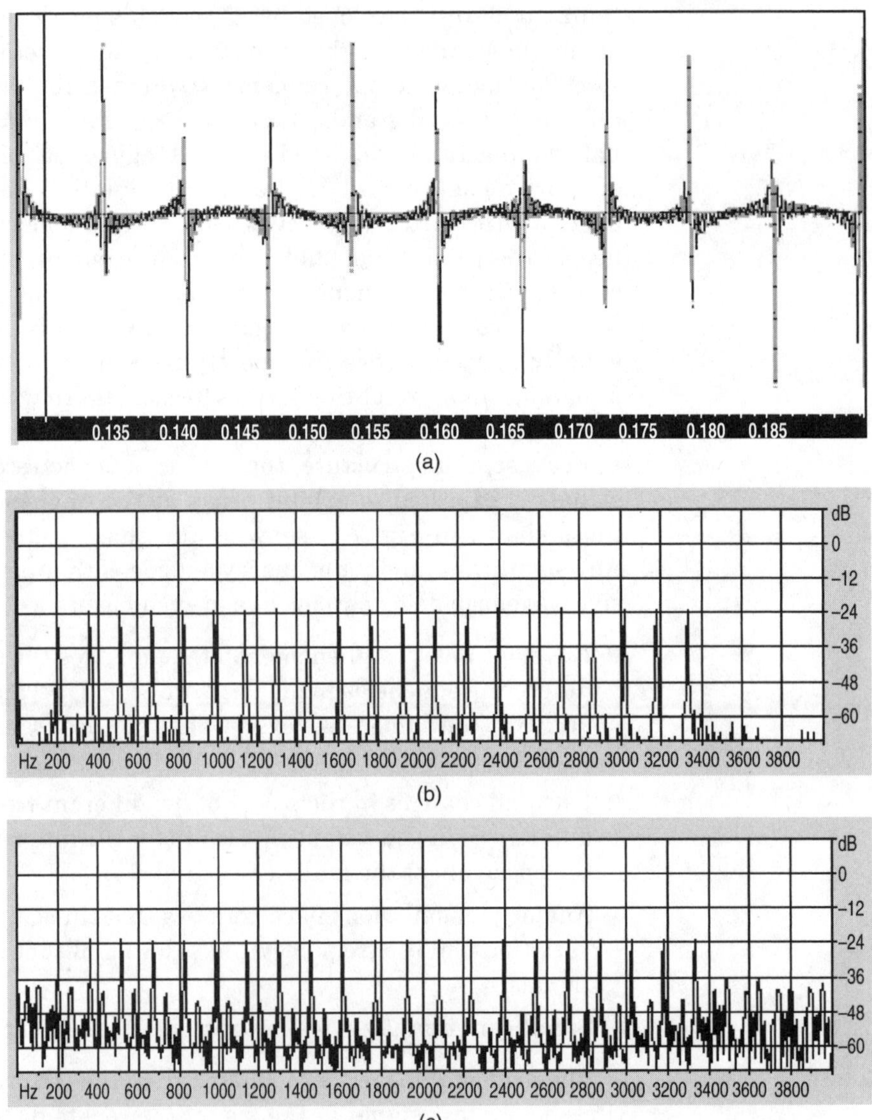

Figure 7-8
21-tone signal characteristics. (*a*) 21-tone waveform; (*b*) clean spectrum–64 kbit/s PCM encoding; (*c*) spectrum with distortion.

In applications of this measure it was discovered that there were substantial differences in the signal-to-distortion ratios for different PCM codecs like those shown in Table 7-3, showing that the differences in signal-to-distortion ratios decrease uniformly with the codec sampling rate. Since lower sampling rates among PCM-encoding techniques are known to degrade perceived quality of voice signals, such results suggested that

TABLE 7-3

Signal-to-Distortion Ratios Measured with 23-Tone Multifrequency Test

Codec Type		
Speed, kbit/s	**Encoding**	**Characteristic Signal-to-Distortion Ratio, dB**
64	PCM	38
40	ADPCM	26
32	ADPCM	22
24	ADPCM	16
16	ADPCM	10
8	CELP	2–10

the signal-to-distortion ratio calculated from received images of multifrequency test signals might be a good basis for a user-perception model.

The early promise of this technique was, however, quickly quashed by subsequent attempts to apply the multitone tests to voice connections encoded with CELP codecs. These tests resulted in extremely low signal-to-distortion ratios in the ranges shown in Table 7-3, although users were reporting very little incidence of noticeable speech distortion over connections tested. The inescapable conclusion was that whatever was being captured by the signal-to-distortion ratio was neither a ubiquitous nor a consistent influence on user perception of voice quality.

This is not to say, however, that the effort was a vain one. Just the fact that a test that had been shown to be a very reliable predictor of how well fax and modem signals would be carried over different codecs failed to predict voice quality revealed a lot about what might be a good predictor. For example, it was surmised that what was causing the 23-tone test to produce such poor signal-to-distortion values for CELP-encoded transmissions (see App. A) was the amplitude jitter resulting from the regeneration of the signals. The low signal-to-distortion values thus correctly predicted that the CELP encoding would confuse the demodulation of signals in which both phase and amplitude were being changed to encode data.

At the same time, our very short course in the physiology of speech and hearing suggests that a measure sensitive to amplitude jitter should not be expected to serve as a good indicator of likely human perception of voice quality. If, as we concluded in that discussion, what is important in phoneme recognition are the relative positions of high values of the spectrum of the sound waves, then it is easy to see how the signal-to-distortion ratio might not correlate with user perception of quality. Frequencies excited by sound waves are not materially affected by variations in amplitude as long as the same basal signal is repeated during the dwell time for a phoneme. Thus, the amplitude jitter that is so destructive to modem signals cannot be expected to have a similarly destructive effect on voice signals.

A Speech Processor The natural first step from the pure tone, multifrequency tests that do not work for human speech signals to something that might is the substitution of recordings of actual speech with known characteristics for the multifrequency tone files. A test procedure like this, based on use of speech as the test signal and execution of analysis of spectral differences to produce the objective measure, was described in a patent issued to George J. Boggs of GTE Laboratories Incorporated in 1989 as follows:

> A method of evaluating the quality of speech in a voice communication system is used in a speech processor. A digital file of undistorted speech representative of a speech standard for a voice communication system is recorded. A sample file of possibly distorted speech carried by said voice communication system is also recorded. The file of standard speech and the file of possibly distorted speech are passed through a set of critical band filters to provide power spectra which include distorted-standard speech pairs. A variance-covariance matrix is calculated from said pairs, and a Mahalanobis D.sup.2 calculation is performed on said matrix, yielding D.sup.2 data which represents an estimation of the quality of speech in the sample file. (U.S. Patent No. 4,860,360, A Method of Evaluating Speech)

This patent (see App. D) goes on to describe an envisioned process whereby samples of original and transmitted speech would be compared to produce the measure described in the abstract and graded in subjective tests to determine the correlation between the measure and the user perception of speech distortion. This particular technique has neither made it into the standards bodies nor, to my knowledge, been offered for inclusion for use in telecommunications test devices. It is probably safe to surmise, then, that the development efforts met with difficulties in obtaining a consistent correlation between the measure and the user perception of speech distortion. If so, such a result might be expected, because, as noted in our very short course in speech and hearing, it is the pattern of sound frequencies rather than specific values that is discriminated in hearing. Consequently, a spectrally weighted measure like the one suggested is not likely to exhibit any correlation with speech distortion as the testing moves from speaker to speaker.

Psychoacoustic Standards

The multifrequency tone tests and speech processor just described, then, illustrate the essence of electroacoustical measures. Such measures assign values to characteristics of the electrically transmitted signals that are independent of any consideration of the content of the waveform being transmitted. There is neither a presumption regarding the type of speaker nor any consideration as to whether the sound frequencies being measured are being generated as part of a particular type of phoneme.

A further refinement of the idea of transmitting a recording of natural speech and comparing the received version to gauge likely user perception of voice quality is, then, to

1. Construct test signals to be transmitted that comprise phoneme sequences interspersed with periods of silence constituting a sample of basal waveforms whose shape and incidence approximate those of the language of the users.
2. Define measures of the differences between the transmitted and received signals that are weighted to reflect variations in the expected effects that different kinds of waveform deformations will have on the characteristics of sound waves of most importance to human hearing.

Perceptual Speech Quality Measure One of the earliest refinements of this type resulted in the Perceptual Speech Quality Measure (PSQM) distributed as ITU-T Recommendation P.861 (Ref. 4) for measuring the quality of speech codecs. The purpose of the recommendation was to set standards for objectively gauging the likely user perception of quality of voice carried via different codecs under a variety of conditions. The envisioned procedure was to use artificial test signals designed to typify natural speech in the way described in step 1; transmit these under conditions that were controlled with respect to the experimental parameters, such as listening level and environmental noise; and calculate a measure with the characteristics in step 2 that could be shown to correlate with the results of subjective user tests.

At the core of this process was to be a model that would "mimic sound perception of subjects in real-life situations" to produce a quality measure that would reflect the different ways that human perception of speech are affected by deformations of timing, frequency, and amplitude of speech waveforms. The model used was based largely on work by Beerends, a scientist with KPN (Koninklijke PTT Nederland), the Royal Netherlands postal, telegraph, and telephone service whose studies and patents are shown in App. D.

To gloss over a highly complex, complicated process, suffice it to say that the ITU recommendation describes a four-step process for the estimation of likely user perception of quality of voice comprising:

1. *Application of a perceptual model.* The PSQM measure is calculated from a comparison of *source* waveforms recorded in the clear, with *test* waveforms that have been encoded and decoded through the codec being tested. The first step in this comparison is the application of perceptual transforms to both the source and test waveforms. These transforms take inputs comprising speech sound waves in the form that they are presented to a telephone mouthpiece and apply algorithms that determine what would be perceived by a human ear listening at the distant earpiece. There are two steps involved. In the first step both the source and test inputs are transformed into speech waveforms that would result from the frequency filtering and shaping that occurs at the send and receive sides of a telephone handset, together with the addition of noise. The outputs, representing the waveforms as they would be presented at the earpiece, are then processed through a model of hearing to create their *internal representations,* comprising the sound information as it would be sensed by the human ear.

2. *Calculation of the difference between internal representations.* The next step is a comparison of the internal representations of the source and test waveforms to identify possibly significant differences in perception of the transformed source and test waveforms. Consonant with the implications from our short course in speech and hearing, the principal differences considered in this process are between the frequency spectra of the source and test waveforms. They are, however, calculated in such a manner as to require preservation of the patterns of distribution of energy, rather than preservation of exact energy levels in exact frequencies.

3. *Application of a cognitive model.* The next step is to interpret the differences between the transformed source and test waveforms to characterize their overall impact in a way that is relevant to user perception of voice quality. In the PSQM system, all operations that cannot be performed on the source signal alone or on the coded signal alone are defined as *cognitive operations.* They are effected in the *cognitive model,* which comprises algorithms that assign different weights to different types of differences between the internal representations of the source and test waveforms. The output from this cognitive model is the PSQM value for the test, representing a combination of the observed differences in which each has been weighted by its expected relative impact on user perception of quality.

4. *Assignment of associated estimates of subjective measures of voice quality.* As a measure of likely subjective perception of quality, the PSQM measure by itself can serve as a tool for analysis of relative quality. Direct comparison of PSQM values may, therefore, be useful in such applications as comparing the performance of two difference codecs, or optimizing the use of codecs, by determining the configuration values that yield the best quality under expected circumstances of use. Conversion of the PSQM into a tool for absolute characterization of voice quality, however, requires the final step of developing a means of producing from the PSQM value an estimate of a measure that reflects likely user perception of quality. The procedure for developing the means to do this recommended in ITU-T P.861 is to conduct subjective listening tests to determine the correlation between the PSQM produced under various test conditions and the MOS. The recommendation explicitly recognizes, however, that this procedure poses a problem:

> Since the relationship between the MOS and PSQM values is not necessarily the same for different languages or even for different subjective tests within a language, it is difficult to determine a unique function which transforms the PSQM value to the estimated MOS value. In practice, therefore, it is necessary to derive such transformation functions for individual languages and individual subjective tests in advance.

Perceptual Analysis Measurement System The endorsement of the PSQM as a standard for the objective measurement of voice quality for telephone band codecs was based on a study of its performance versus competing techniques by the names like LPC Cepstrum Distance Measure, Information Index, Coherence Function, and Expert Pattern Recognition. A psychoacoustic measurement technique very similar to the PSQM that was not considered in these competitions was the Perceptual Analysis Measurement System (PAMS). This system was developed independently by the PTT (postal, telegraph, and telephone agency) for the United Kingdom, later known to the world as British Telecom (BT), for internal use. The principal developer was a BT scientist, Mike Hollier. His relevant patents are summarized in App. D.

As indicated, the PAMS is, in concept, very much like the PSQM, following the same general approach. The basic procedure involves transmission of a source signal and capture of a test signal comprising the received image of that source signal at the far end of the kind of voice connection being tested. Comparison of the source and test signal is effected by first transforming both through an *auditory transform* that maps sound waves into their equivalent representations as they would be heard. The result of the PAMS auditory transform is called a *sensation surface.* It comprises a description of the flow of the source and test signals represented by calculation of the frequency spectrum, at equally spaced, short-time intervals across the sample. The distortion resulting from transmission of the source signal is then gauged by subtracting the source sensation surface from the test sensation to produce an *error surface.* The basis for objective measurement of the distortion then becomes calculation of the average of different varieties of differences exhibited in the error surface, such as positive and negative values occurring in the error surface at different frequency equivalents in the sensation surfaces. The processing of the identified components of the error surface to produce estimates of subjective measures of voice quality is then accomplished by a formula that was developed by using regression techniques to fit test results developed from extensive in-lab subjective testing.

As seen from the preceding description, the basic PAMS measurement approach differs from the PSQM only in the details of the methodology. Both utilize transforms that purport to produce measures based on a comparison of the source and test signals as they would be heard, rather than the way that they would be represented in an electroacoustic waveform like that shown in Fig. 7-8. And, both rely on subjective user testing under controlled conditions to establish the relationship between the measures calculated and the predicted MOS.

There were, however, some significant differences in implementation of the PAMS, including:

1. *Use of specially designed source waveforms.* The waveform processing in either the PAMS or PSQM might be applied to almost anything that could be asserted to typify conversational speech waveforms in the language to be studied. However, the BT scientists invested a lot of time in developing a proprietary artificial speechlike test stimulus (ASTS). The various versions of this test waveform contained particular collections and conjunctions of phonemes that not only typified human speech, but surfaced particular second-order deformations whose impact on user perception of speech quality had been gauged through extensive subjective tests. The idea in using such waveforms was to capitalize on the ability to use computers to digitally construct almost any conceivable waveform to create ones whose use could automatically reduce some of the computational complexity of psychoacoustic measurement.

2. *Production of two measures of voice quality.* In addition to producing an estimate of a MOS for listening quality, the subjective testing also sampled and established formulas for predicting a MOS for listening effort. The description of guidance for assigning opinion scores for listening quality and listening effort are shown in Table 7-4. As was pointed out in our discussion of the basics, the quantification of two types of opinion, rather than quality alone, at least moves us in the direction of making the distinction between the quality of what was heard and usability of the connection that was found to be so important in the service attribute testing.

3. *Procedures for checking and realigning the source and test signals.* As should be clear from the way that the PAMS and PSQM measurements are calculated, any loss of synchronization in the comparison of the source and test signals will result in invalid measurements. As work with the PAMS continued to attempt its application to packet-switched voice services, however, it was discovered that there was in this environment a potential for significant differences in the timing of the source and test signals. The most obvious source was adjustments in the jitter buffer created by adaptive jitter controls. The adjustments would be applied in periods of silence, creating an offset between the time that the next voice signal started in the source and test signals. Since the use of adaptive jitter buffers was expected to be widespread, the potential for this kind of loss of synchronization produced a real threat to the viability of use of either measurement scheme for gauging voice quality in operational networks.

TABLE 7-4

Listening Quality and Listening Effort Criteria and Scores

Quality	Effort Required to Understand Meanings of Sentences	Score
Excellent	Complete relaxation; no effort required	5
Good	Attention necessary; no appreciable effort required	4
Fair	Moderate effort required	3
Poor	Considerable effort required	2
Unsatisfactory/bad	No meaning understood with any feasible effort	1

Perceptual Evaluation of Speech Quality The PAMS was adopted for applications within BT in the early 1990s and first demonstrated for application to VoIP in 1998. The PSQM was investigated by the ITU as early as in the 1980s and was recommended as a standard for testing codecs in 1996, which was later updated early in 1998 to reflect improvements that were wrought through continuing development. In February of 2001 the improvements and refinements of both PSQM and PAMS were integrated to produce a new standard, described in ITU-T Recommendation P.862 (Ref. 5). The new system was called the Perceptual Evaluation of Speech Quality (PESQ) and described as "an objective method for end-to-end assessment of narrow-band telephone networks and speech codecs."

The new PESQ promulgated in ITU-T P.862 was, in fact, a combination of the perceptual models developed for the PSQM, as revised and improved through 1999, and the processes for detecting and correcting losses of synchronization between the source and test signal developed for the PAMS. The representation of the contributions to it from the references was completely balanced, citing three articles by Beerends of KPN and three by Hollier. Moreover, as shown in App. D, the system exploited intellectual property rights from both KPN and BT. (*Note:* Michael Hollier later secured licenses from KPN and BT which enabled him to found a company, Psytechnics, Inc., whose purpose is the distribution of software and technology that exploit the patents that protect components of the PAMS, PSQM, and PESQ. A wealth of information on the whole realm of psychoacoustic measurement of voice quality can be found on the Psytechnics website at *www.psytechnics.com.*)

Limitations Although the incorporation of the PAMS methodology for handling variability in the synchronization of source and test signals enabled the application of the PESQ to packet-switched connections, the approach taken in the psychoacoustic test procedures described here have four characteristics in common that limit their applicability.

1. *Requirements for level alignment.* In order to measure the difference between the source and test signals, both the PSQM and PAMS methodologies require an initial step of compensating for any signal power attenuation occurring over the connection being tested. The reason is that the waveform distortion measurements must reflect the differences between the source and test signals at the levels at which the source signal was originally recorded. This requirement, in effect, eliminates from the psychoacoustic assessment of user perception of quality any consideration of the power of the speech signals received, which determines the likely incidence of reports of "low volume" in the service attribute test.

2. *Combination of effects.* In the perceptual models used for either the PSQM or PAMS the principal measurements reflect a difference between what is heard and what was transmitted. In applications outside of the tightly controlled environment of a test laboratory, these differences will depend both on waveform deformations and on noise that is added to the test signal during the course of its transmission. As a consequence, the differences measured and gauged via the psychoacoustic measurement techniques described here inextricably reflect the combined effects of noise and other kinds of waveform distortions, such as those caused by high bit error rates on a digital link or by dropped frames in a packet-switched voice service. In practical terms, this means that when the PAMS or PESQ produces a low estimated MOS, there is no simple way to determine the source of the deterioration.

3. *Comparisons of received and transmitted signals.* Perhaps the greatest weakness in the psychoacoustic measurement techniques that have been described here is that all measurements are based on a comparison of a source and test signal. This means that the measurements cannot reflect the effects of either echo, which is a major impairment that may affect the user perception of quality of all kinds of telephonic communications, or the round-trip delay, which becomes a concern in satellite and packet-switched communications. The reason, of course, is that echo is a phenomenon principally experienced by talkers. There are cases in which the echo is loud enough to produce an audible return back to the destination, but such instances of listener echo are rare in today's systems. Since echo from the source signal would be most likely to be perceptible at the point of transmission of the source signal, this means that neither the PAMS nor the PESQ by themselves can reflect user reaction to echo. Similarly, since the source and test signals must be synchronized in order to calculate the differences, there is nothing in either the PAMS or PESQ measures that can reflect the disruption of conversational rhythms caused by excessive round-trip delay.

The limitations cited here are not revelations. They are, in fact, fully recognized and described in the caveat in ITU-T P.862, paragraph 4, (Ref. 5) which states, in part:

> It should also be noted that the PESQ algorithm does not provide a comprehensive evaluation of transmission quality. It only measures the effects

of one-way speech distortion and noise on speech quality. The effects of loudness loss, delay, sidetone, echo, and other impairments related to two-way interaction (e.g., center clipper) are not reflected in the PESQ scores. Therefore, it is possible to have high PESQ scores, yet poor quality of the connection overall.

4. Another, more subtle limitation that may be a revelation to some, however, is that in a packet-switched network, *a single value of a measure like those calculated in these test procedures cannot be meaningfully interpreted.* The problem is this. In a circuit-switched network the conditions that degrade quality, such as noise or a high bit error rate, persist for the duration of the call, producing effects that are equally distributed among all parts of the waveform sampled. In packet-switched networks the principal contributor to speech distortion is dropped frames, which are at best randomly distributed discrete events and may, in fact, be manifested as groups of dropped frames or short periods in which many are dropped. Because of this, the pattern of dropped frames in any particular sample may be one in which the effects of dropped frames are not equally distributed between periods of speech and periods of silence. Since the psychoacoustic measurements described here all ascribe a different weight to disturbances of speech and disturbances of silence, this means that the same underlying conditions can, as illustrated in Fig. 7-9, result in substantially different measures for the same sample. In practical terms this means that evaluation of voice quality of a particular service by analysis of these measures must be based on a repeated sample that is large enough to produce an accurate estimate of the average of the expected variations in effects.

Limitations like these do not necessarily vitiate the usefulness of active psychoacoustic testing of voice quality. In digitized voice services that are originated on four-wire connections and packet-switched end to end, loss, noise, and echo should not be a problem as long as the service has been designed and validated to meet standards for these factors on installation. Changes in such measures in this environment observed in large samples can, therefore, be reliable indicators of the emergence of problems that are degrading the likely user perception of voice quality. Comparisons of large samples of the measures for different services can similarly be an effective means of verifying achievement of expected voice quality.

What these limitations do imply, however, is that none of the P-family of psychoacoustic measures (PSQM, PAMS, PESQ) should be used as a stand-alone gauge of quality of packet-switched voice services.

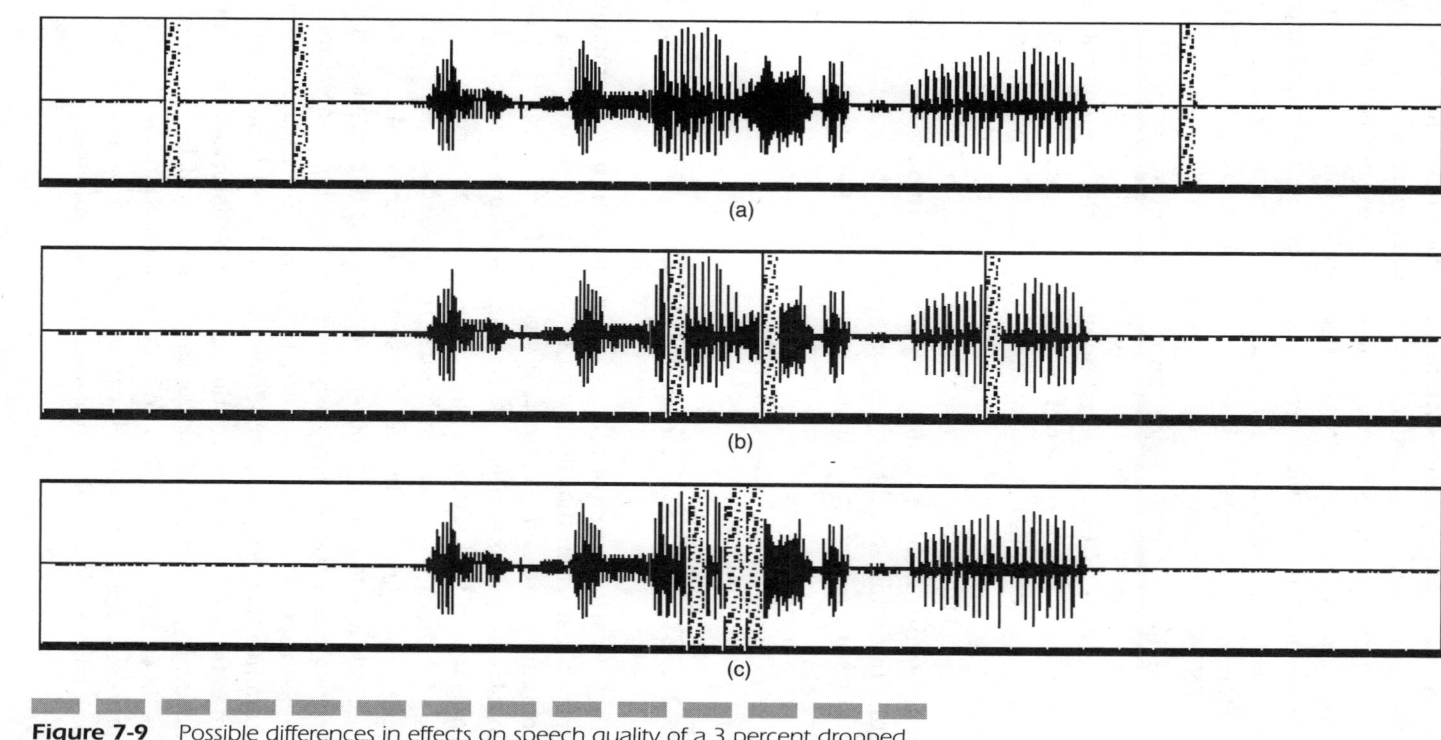

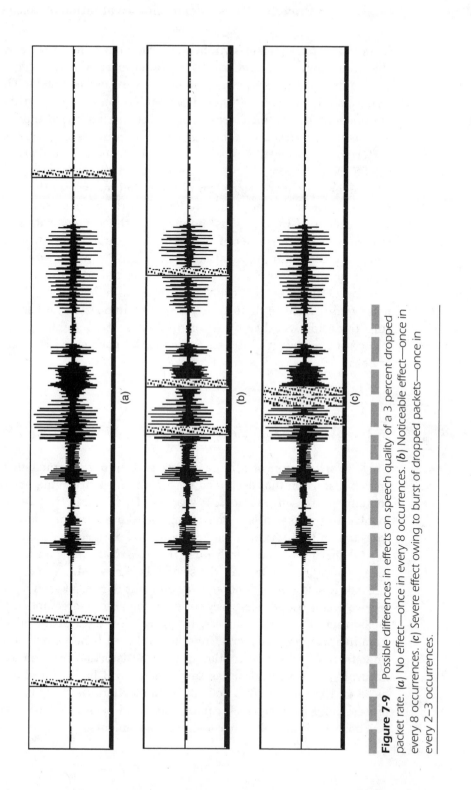

Figure 7-9 Possible differences in effects on speech quality of a 3 percent dropped packet rate. (*a*) No effect—once in every 8 occurrences. (*b*) Noticeable effect—once in every 8 occurrences. (*c*) Severe effect owing to burst of dropped packets—once in every 2–3 occurrences.

Even in the pure packet-switched environment where they can be reliable indicators of emergence of problems, particularly those attributable to dropped packets, these measures are not complete. The monitoring and assessment of likely user perception of quality still must be complemented with measures that reflect the likelihood that users are experiencing round-trip delays great enough to disrupt conversational patterns. Moreover, in any services in which the calls are originated in the PSTN and passed into a packet-switched network for onward delivery, the quality of the voice can be degraded by exactly the effects previously described as not being reflected in the measures. In the terms introduced in Chap. 3, this means that these measures are simply inadequate for purposes of assessing the quality of voice in hybrid transport networks. Application of the PESQ may, for example, reveal a poor predicted MOS for a particular end-to-end call across a hybrid transport. However, because the MOS value reflects the effects of both noise and waveform distortions, it will be impossible to use that measure to decide whether the poor quality is attributable to the performance of the circuit-switched network, the packet-switched transport, or both.

Passive Measurement Techniques

Another limitation of all the active techniques for measurement of speech distortion just described is that the measurements are based on comparison of source and test signals. This means that testing must involve transmission of a source signal whose structure is known and collection of that signal at the receiving end of the connection being tested. All testing of this kind must be *intrusive,* based on allowing the tester to originate calls over the service being tested.

Since it is not always possible or efficient to rely on such access to the service being tested, it is very useful to be able to accomplish the same measurements *nonintrusively,* by processing data acquired passively from live telephone conversations. To do this, it is necessary to be able to infer the likely user perception of speech distortion from an analysis of natural speech waveforms as they are, or will be, received at the earpiece on the far side of a telephone connection. As described next, the processes for creating such capabilities have been developed using both psychoacoustic and electroacoustic measurements.

Psychoacoustic-PAMS Extensions

The best current examples of the psychoacoustic implementation of measures designed to process natural speech waveforms are various extensions of the PAMS technology to applications in nonintrusive testing. The ones that have been announced by dint of their being patented include:

1. *A method of reconstructing speech content from the intercepted speech waveforms and effecting comparisons.* This technique is described in U.S. Patent 5,848,384, Analysis of Audio Quality Using Speech Recognition and Synthesis. The idea was to enable the equivalent of the psychoacoustic measurement by analyzing a captured test signal to reconstruct the most likely undistorted version of the source signal and by applying the measurements just as if the reconstructed signal had been the transmitted source.

2. *A similar method based on identification of talker invariant characteristics.* In U.S. Patent 5,940,792, Nonintrusive Testing of Telecommunication Speech by Determining Deviations from Invariant Characteristics or Relationships, the idea in example 1 is simplified by searching the natural speech waveform for instances of characteristics that are independent of the speaker and basing the reconstruction of the presumed source signal on those characteristics. Examples of such elemental reconstruction described in the patent are vowel sounds, for which the spectral pattern is consistent, and the level in voiced fricatives, for which the spectral content of the fricative varies with the speech power of the speaker. This method of measurement, then, is implemented by looking at short segments of the captured speech signal and presuming that matches to vowel sound and fricative patterns were the result of instances of transmission of the underlying "pure" patterns.

3. *A complex method based on a model of vocal articulation.* The formal description of this technique, taken from U.S. Patent 6,035,270, Trained Artificial Neural Networks Using an Imperfect Vocal Tract Model for Assessment of Speech Signal Quality, is

> A non-intrusive method of assessing the quality of a first signal carrying speech, said method comprising the steps of: analyzing said signal carrying speech to generate output parameters according to a spectral representation imperfect vocal tract model capable of generating coefficients that can parametrically represent both speech and distortion signal elements, and

> weighting the output parameters according to a network definition function to generate an output derived from the weighted output parameters, the network definition function being generated using a trainable process, using well conditioned and/or ill-conditioned samples of a test signal, modeled by imperfect the [*sic*] vocal tract model.

A less formal description of all these techniques is that the objective is to sample a little piece of a captured speech waveform to determine whether it could have been articulated by a human speaker; identify its content and compare it to the ideal waveform from the presumed speaker to measure distortion; and proceed to the next segment. The basic concepts here are workable, but the amount of processing involved is certainly daunting.

An Electroacoustic Technique

In contrast to the inherent complexity of trying to retrofit psychoacoustic measurement techniques for use in nonintrusive testing, there is a simple premise that supports electroacoustic measurements of natural speech waveforms to produce indicators of likely user perception of speech distortion. The premise, described in U.S. Patent 6,246,978, Method and System for Measurement of Speech Distortion from Samples of Telephonic Voice Signals, is that what is reported as "speech distortion" in service attribute tests is the *presence of changes in the waveforms that could not have been articulated in human speech.* Such changes include, for example, the power spikes in "extra" frequencies that appear when quantizing noise is not masked by line noise. In this case, what is described as "raspiness" in the voice of the speaker is, in fact, due to local changes in the shape of the smoothly articulated waveform that are faster than those that could have been voiced.

More generally, examination of the possible changes in voice signals in light of the mechanics of speech described in our short course at the beginning of this chapter supports the intuitively evident hypothesis that *the physiology of articulation naturally bounds the rate at which the amplitude of a natural speech waveform can change.* If this is so, then it follows that at least some of the conditions that will result in reports of the presence of speech distortion will be reflected in the mathematical characteristics of the speech waveforms in time, rather than in the frequency spectra that are the principal concerns of psychoacoustic measurement. In particular, this implies that the incidence of abnormally fast changes in the amplitude of waveforms captured at the receiving end of a voice connection should materially affect the reported incidence and severity of "speech distortion" as it is described for SAT subjects.

In mathematical terms, changes in the amplitude of a waveform in the time domain are reflected in the first derivative of the amplitude as a function of time, describing the *speed of changes,* and in the second derivative, representing the local *acceleration* of amplitude. These characteristics of continuous sound waveforms can be readily estimated from the discrete samples of amplitude obtained from PCM digitization as the first and second differences of consecutive samples. That is, if A_0, A_1, and A_2 are three consecutive samples taken at fixed time points, 0, t, and $2t$, then the first differences are $(A_1 - A_0)$ and $(A_2 - A_1)$. The second difference is $(A_2 - A_1) - (A_1 - A_0)$.

The hypothesis that the speed and acceleration of changes in amplitude in natural speech signals are strongly bounded then suggests a family of measures based on the distribution of such first and second differences observed in digitally encoded speech waveforms. One of the measures in that family that was found to consistently exhibit a very high correlation with user reports of the incidence and severity of speech distortion as reported in SATs was the ***kurtosis*** of the *second differences* of amplitudes of natural voice waveforms. The kurtosis of a sample is a statistical measure of the tendency of the observed values to cluster about the mean. A normal distribution, which would represent the shape of the distribution of a sample of random line noise, has a kurtosis of 3.0. The kurtosis of the amplitudes from natural speech samples can be shown to have much higher values, which is the expected result if the incidence of very fast changes in or acceleration of amplitudes is very small in natural speech.

This measure has historically been favored in applications to voice quality analysis for MCI/WorldCom for three reasons:

1. *It has high inherent credibility.* As seen from the description here, this measure is intuitively pleasing, because it is based on a very simple theory compared to those on which psychoacoustic measurements are based.

2. *It is easy to calculate.* Since the measurement is based on what happens in the time domain, it can be calculated without invoking the computationally intensive Fourier transforms that are necessary to produce the frequency spectra that become the basis for psychoacoustic measurements.

3. *It can be used both in passive and active testing.* Because the evaluation of voice quality is based on data derived from digital images of natural speech waveforms, this measurement technique can be readily applied to voice waveforms captured via nonintrusive data acquisition. However, since the only stimulus is undistorted natural speech, the measurement can be applied in active test setups by transmitting and analyzing readily encoded speech samples. Moreover, the use of prerecorded speech samples for the source signal creates an ability to tailor what is sent to ensure that only voice waveforms are represented in the source signal and to verify in advance that the measure will indicate acceptable quality, given that there are no significant distortions resulting from its transmission.

Another characteristic of these electroacoustic techniques for measuring speech distortion that makes them very useful for applications in evaluation of packet-switched voice services is the fact that something like the kurtosis of the second differences of digitized speech waveforms is *not greatly affected by dropped frames.* Unconcealed dropped packets create nulls in the received speech waveforms that must be excluded from the calculation of the second differences, since only speech waveforms are supposed to be sampled. Concealed packet losses result in at most two excessive jumps in amplitude per 100 data points. Thus, a packet loss of even 5 percent adds at most 0.1 percent to the distribution of values that reduce the kurtosis.

In practical terms, this means that these readily calculated measures can be used in conjunction with measures that exhibit high correlation with dropped frame rates across a packet-switched network to infer both the incidence and the source of speech distortion in a hybrid transport service.

Comment

The description of techniques for measuring speech distortion here are intended to be illustrative rather than comprehensive. There are, in fact, numerous refinements to each of the basic techniques that have not been elaborated. There are also, in all probability, a large number of confidential ongoing efforts in this area, particularly in the pursuit of psychoacoustic measurements, that may later see the light of day and revolutionize the technology.

CHAPTER 8

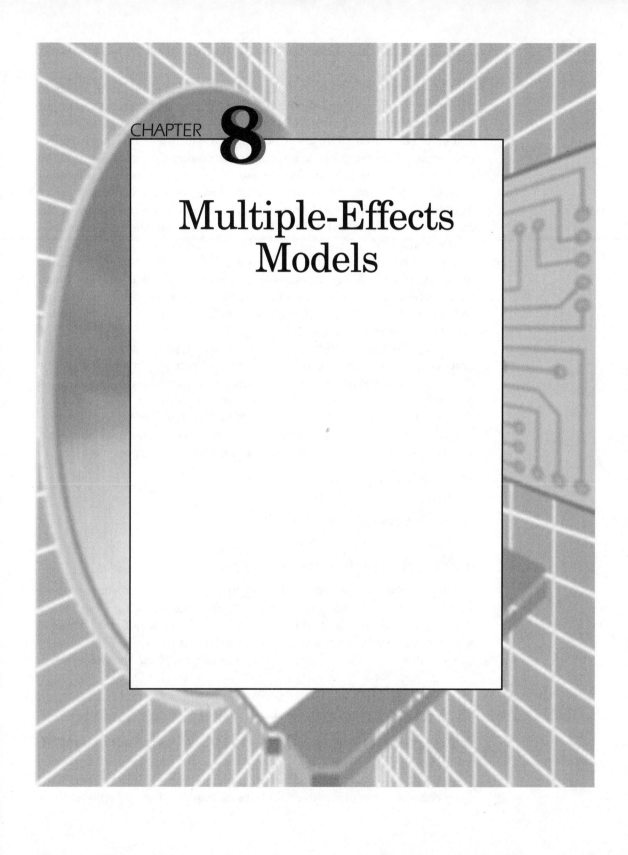

Multiple-Effects Models

The broad objective of the modeling efforts described in Chap. 7 was to produce measures of the deformation of speech waveforms that might serve as reliable indicators of the likely user perception of quality of voice. As seen from their descriptions, however, these measures reflect only some of the effects of telephone transmission that may degrade the quality of what is heard. The measures reflect the overall effects of waveform deformations caused by such things as signal filtering, codec encoding, errors in digital transmission, dropped packets, and, in the case of the psychoacoustic measures, noise. However, they are completely invariant with respect to other characteristics, such as low volume and echo, that have been demonstrated in SATs to affect user perception of quality of what is heard. Moreover, because those measures are calculated independently of the average round-trip delay, they can give no indication of possible degradation of usability of service from disruption of normal conversational patterns.

As suggested earlier, this is not a revelation. However, it does imply that none of the speech distortion measurements described in Chap. 7 can by itself be expected to be an adequate basis for gauging likely user perception of quality of a voice service. Rather, what is needed is a user-perception model that will produce measures reflecting the combined effects of performance characteristics that determine the incidence and severity of at least five different manifestations that may be perceived by telephone users:

- Low volume
- Noise
- Speech distortion
- Echo
- Effects of round-trip delay

To appreciate all the factors that might come into play in shaping those manifestations, we recommend that you read the following story, which recounts the adventures of three phonemes in "Telephone Land." Those of you who prefer your information to be imparted without an icing of whimsey can skip directly to the next major section heading or skim what follows.

A Modern Fable—The Three Little Phonemes and the Big, Bad Network

It seems one day in Phoneme Heaven (which is where all these short-lived little critters go as their reward for helping human communications) three phonemes happened to get together as they were just lying around together in the soothing wash of the Great Eternal Reverberation which would eventually recall them at some time for rebirth as yet another sound. As they lay there, just damping out, they began to recount their life experiences, which were unusual, because these particular phonemes had been transmitted over great distances through the always exciting, sometimes terrifying ordeal of travel through Telephone Land.

The first little phoneme's story went something like this, assisted by Fig. 8-1 (which, of course, could not have been drawn or referenced, because they were, after all, but transient little phonemes, and dead ones at that. The nice thing about being in Phoneme Heaven, however, is that it is a place where that which is possible, no matter how improbable, is possible).

First Little Phoneme

"I was born at point A as a beautifully articulated 'oh' uttered by a man with a very pleasant deep, resonant baritone voice. I had hopes that such a wonderful start would mean that I was destined to be a beautifully sung note. Alas, however, such was not the case, and I knew it quickly, because my dwell time was just too short to be musical.

"What happened next was even more demeaning. No sooner had I been born at point A, I was transformed into an electrical signal and beat up something fierce. By the time I got to point B, my basal shape had been filtered, so all that my spectrum had left were frequencies between 300 and 3400 Hz. That wasn't so bad, because I was still a very clear 'oh' that no human would have trouble recognizing. But still, it did smart a little bit to know that I had lost some of that beautiful resonance.

"I am proud to say, however, that I was, at that point, still a nicely powered electrical signal, running at about −17 dBm, which is just about average for

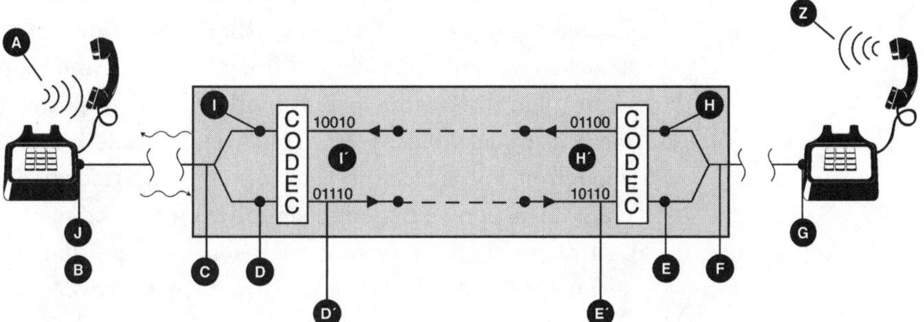

Figure 8-1 What the first little phoneme saw.

us. [*Author's note:* dBm is a measure of signal power, expressed in decibels relative to 1 milliwatt.] What happened next, though, was that I traveled along this wire that was also the one carrying some speech from the other end of the connection. The other and I didn't get in each other's way, but we sure could have, traveling so close together like that. As I traveled along that line, though, I lost some of that nice power and got a few pricks and bumps caused by some line noise riders that latched on along the way.

"By and by I got to point C, by which time I had lost 3 dB of that original power, which I am told is considered to be just about par for this course. I have heard some of the other phonemes tell of cases where the loss is 2 to 3 dB more, so I guess that I should be thankful, even though others don't seem to lose so much power.

"Anyway, I got to where the wire ran into a building with a whole bunch of other wires, and at point C, I was whipped around this strange device. Someone told me once that this thing is called a *hybrid,* and that it can separate two signals going in two directions over the same line, or put two separate signals together. Whatever it was, it did a nice job, because I was suddenly on my own line, no longer competing with those other signals from the opposite direction.

"However, that didn't last long. My signal was almost immediately thrown into a device at point D, and my shape, which once at least resembled the sound I was born as, was converted into pulse waves representing strings of 1s and 0s flowing out at point D′. Now, it is hard to keep any sort of phonetic consciousness when you've been broken to bits like this, but I am almost sure that I maintained a good standing in whatever happened, because when I regained my senses after having gone to the other side of wherever I went, arriving at point E′, and coming out at point E again looking like my old self, I didn't notice any change in power at all, and I didn't seem to have picked up many more noise riders. There were a few strange dinks here and there, and a little bit of error in my amplitudes, but all and all it didn't seem to be too bad...

"...at least as compared to what happened to me at point E. When I got there, my electrical shape was pounded down a whole 6 dB and I was immediately thrown into the other side of one of those hybrid things, smashed onto another shared line at point F, and zipped down to point G, where I arrived after losing another 3 dB of power and picking up still more line noise riders. It was a wonder that I came out of all this with enough of my old shape that when I was reformed as a sound wave at point Z the person listening actually understood me to be 'oh.'

"Having checked around with some of the other phonemes up here, it seems that at least part of what happened between points E and F was on purpose, to help with other kinds of things that can happen. Moreover, I am happy to report without too much modesty that the 12-dB loss in power that I experienced in getting to point G is *exactly* what those humans who built Telephone Land want to happen.

"And, I think I know why. You see, there was one other thing that happened to me that I am ashamed to admit. When I got thrown into that hybrid at point E, I didn't just get pounded down. I also went schizophrenic. It was like two of me were created. One of me went on to point G like I said, but the other me was split off and hurled back to point H, through to H′, and eventually all the way back to where the electrical me was first separated out, so that I didn't compete with the signals coming in. Now when that me got to point I, it was pounded down 6 dB, just like what happened back between points E and F. As a consequence, by the time the second me got back to point J it had lost 15 dB from the power that I had when I was running between points D and E. Since the second me that had been created was already but a faint replica of my then and former self, all that extra pounding made sure the me that was split off me didn't have much power when it got back, so it was hard to hear. I hate to think what might have happened if I hadn't been pounded down between points E and F, and the other me hadn't been hit that hard again, back at I.

"Even so, it was all quite an ordeal for this little phoneme, and I am glad that I had the shape and power to stand up to it. All in all a very good life's work against great difficulties, I'd say."

Second Little Phoneme

The first little phoneme looked around at an audience of attentive other little phonemes that had gathered to hear his story, absolutely sure that what it had endured was the odyssey of many, many lifetimes, and certainly not to be bested on that day. You can imagine, then, its chagrin when the end of the story was greeted by a snort.

"Stuff and piffle," declaimed the second little phoneme. "I'll just bet you are proud of yourself. But that's just because you've never had a *really* rough trip across Telephone Land."

"Oh yeah?" said the first little phoneme. "Then suppose you tell us what happened to you that was worse!"

"Gladly," replied the second little phoneme, as it added to the picture to make Fig. 8-2. "How about this?

"I, too, was formed as a perfectly articulated 'oh,' and I was a lot longer than you, since I was intended to draw out more information. You know,

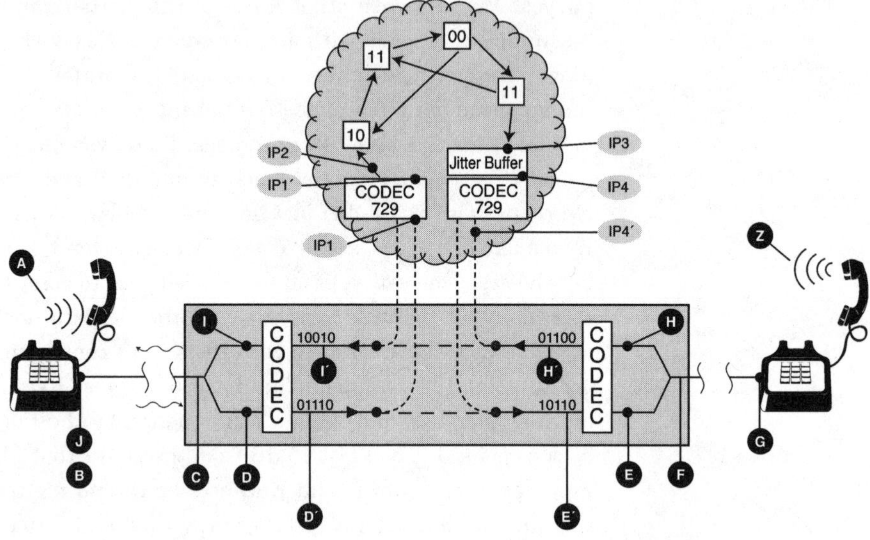

Figure 8-2 What the second little phoneme saw.

when one of those humans says 'ooooh?' instead of saying 'Tell me more.' Anyway, I followed the same path you did until I got to point D′. At that point, as best as I could tell in my digital stupor, *my bits* got thrown into *another digitizing thing* at point IP1. And if you think for one minute that the plain old digital stupor is as bad as it gets, try going one step further, and have your groups of bits that at least represented samples of your amplitudes crunched into numbers representing some sort of compact description of your pieces.

"This is exactly what happened to me. But wait, there's more, because each of those groups of digits describing my pieces was sent out on its own into a maze of paths to find its way to the designated assembly point at IP3. Oh, it was a nightmare! Parts of me were packed into little transports and went this way; other parts of me went that. And my bits and pieces were supposed to be delivered in line at point IP3 in time to be called for reassembly of the strings of bits that came out at point D′ to go to E′.

"Well, you can imagine. Some pieces never made it. Others were not there when they were supposed to be and had to be left behind. And, some of the pieces that made it had errors that really garbaged up my pristine waveform. To help out, sometimes when a piece was not there when needed, the digitizing thing would just clone the last little piece and send it along instead. That was at least better than just dropping the power in that piece to zero in what was fed into the digitizer thing at point E′.

"And *waiting!* You've never seen such waiting as had to be endured in that process. First, there was a wait to get anything going at point IP1, because the digitizing thing had to accumulate 10 ms of my bits to even decide what description to create, and then it took 5 ms to create the description. And, then that description was launched and would get to one point and have to wait before being sent on to another, and on and on. I'm not sure, but I think that I was in that state of disarray a whole lot longer than we were in our digital stupors between points D′ and E′.

"Now I'll admit that there was one part that was a little less disconcerting. The reason is that the part on Telephone Land I was on had a thing called an *echo canceller* sitting across the two sets of wires at points H′ and E′. Just like you, I went schizophrenic at point F with the creation of another me. However, what that thing did was spot something that looked like what went in at point E′ on the line at point H′ and took it out before it could head back toward point J. And that was probably a good thing, because the other me was running so far behind me that I'll bet it would have been cleanly separated from me back at point J, causing some real mischief.

"So there."

Third Little Phoneme

At that point, the first and second little phonemes were clearly ready to launch into a major argument. However, before the first little phoneme could retort to the "So there" challenge, both were completely dumbfounded by a gale of derisive laughter from the third little phoneme that had originally joined them in their reveries.

"And just what is so funny?" asked the first little phoneme. "We've both gone through a terrible ordeal in Telephone Land and barely survived to tell about it. I would think you would show a little respect."

"You guys went through all that stuff, because the human who articulated you was old-fashioned...way behind the times. I've taken a couple of trips through Telephone Land after being articulated by a human using a new kind of telephone, and it's something else again," it said, starting to add to the picture to get Fig. 8-3.

"In those trips, I've started out at point IP-tel, and been immediately run through a digitizer thing that does everything at once, right there. No filtering. No 3-dB loss. No noise riders. And no hybrid. When you get chopped up that quickly, there's no digital stupor. Just a trip in a cloud that's a gas."

"And just where have you ended up on these little Telephone Land trips?" asked the first little phoneme.

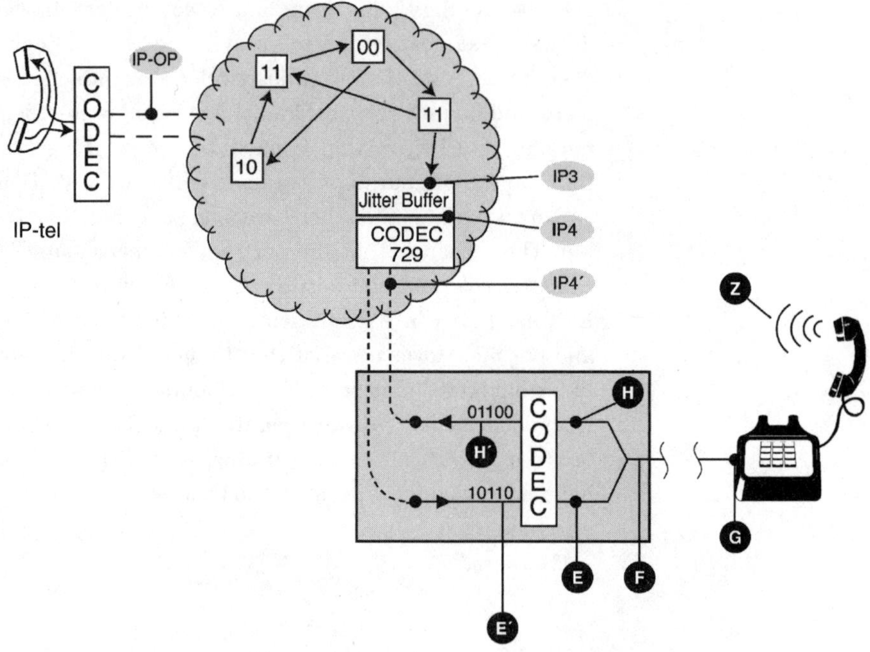

Figure 8-3 What the third little phoneme encountered.

"Right back out into another one of those places like where I started. The pieces don't reassemble until they have gotten to the place where they're going to be used. Again no noise riders, no 3-dB loss, and no filtering. Just a come back from neat trip into the cloud. Oh, I've had some of the problems that you've heard about with waiting and some pieces not getting there in time. But all in all, not much has happened."

At this point the other two little phonemes went into hysterics of their own. However, before the third little phoneme could ask them why, it was called up for rebirth, which, it was announced by the Great Reverberation, just happened to be as a phoneme articulated at an IP-tel point. Knowing this, the other two little phonemes just smiled at each other and waited.

After what seemed to be an inordinately long time in Phoneme Heaven, the third little phoneme returned and took its place next to the other two, but without the audience, which had long since lost interest and drifted off elsewhere to damp out. The third little phoneme was obviously distraught.

"Bad trip?" asked the first little phoneme with an impish smile.

"Yes," sighed the third.

"Un-huh," taunted the second. "Do you want to tell us about it? Or, should we tell you?"

"If you think you can tell me about what I just experienced, just be my guest!"

"Well," said the first little phoneme, "you started out the same way..."

"...but when your pieces got to the reassembly point..." chimed in the second,

"...you came out as a bit stream at point E′ instead of an electrical wave!" they said in unison.

"Then," said the second, "when you got to point E, you came out too loud with no noise riders so that even when you got pounded down 6 dB it wasn't enough and you were a loud raspy, ugly phoneme by the time you got to point Z."

The first joined in. "And, when you got whipped around that hybrid, a large split of you showed up at point H, and the extra power and extra frequencies distorted you to the point that the echo canceller was fooled into thinking that you were speech from point G and missed the other you..."

"...which showed up back at point IP-tel as the worst, most awful sounding echo that had ever been heard by the human who articulated you," they finished in unison.

"Louder," said the first.

"Coulda been down 15 dB, but wasn't," said the second.

"More distinctly separated."

"Coulda been there 300 ms earlier, but wasn't."

"Uglier."

"Mighta had the amplitude errors smoothed out a little by some noise riders, but didn't."

The moral to this story is that packet-switched voice origination will not be compatible with circuit-switched terminations unless the loss plan for the circuit-switched network is recognized and accommodated at the origin.

For the more literal-minded, this little fable also contains the following information:

- On a circuit-switched connection, we expect to see a 12-dB end-to-end loss comprising a nominal loss of 3 dB each on the local subscriber loops that originate and terminate the call and a 6-dB pad at the far end of the connection inserted to mitigate the possible effects of signal power reflection by the hybrid. The local loops are also the major source of ordinary white noise on a telephone connection.

- The digital long-distance transport in the circuit-switched environment is generally 64 kbit/s G.711 standard pulse-code modulation, and the digital transport for voice across a packet-switched network will probably be 8 kbit/s G.729 code-excited linear predictive encoding. The G.729 operates on G.711, so there will always be a first step somewhere in which an electrical analog signal will be encoded into a G.711 PCM signal.
- Digital encoding and transport may result in perceptible deformations of the speech waveforms. The possible causes of such deformations are
  - Quantizing noise representing the difference between the input amplitudes and their encoded values
  - Amplitude clipping resulting from input of waveforms with power greater than the dynamic range of the codec
  - Phonemic clipping caused by application of voice activity detection and silence suppression routines
  - Audible deformations resulting from high bit error rates on the digital transport links
  - The effects of dropped packets in a packet-switched network, principally those attributable to jitter, on the reconstructed speech waveforms
- The incidence and severity of audible echo at the origin of the echoed speech depends on what happens at the distant end of the connection. Specifically, the echo depends on
  - *Echo path loss,* which determines how much speech power reflected by the distant hybrid gets back to the origin. This depends, in turn, on the amount of speech power reflected by the hybrid; the amount of loss deliberately inserted to reduce the level of that speech power; and the effectiveness of any echo cancellation at the far end of the connection.
  - *Echo path delay,* which determines how much the echoed signal is separated in time from its original articulation. Very short delays, less than about 20 ms, do not separate from the original speech enough to be heard. Thereafter, the echo becomes audibly separated from the original speech, with greater delays causing more pronounced echo.

These effects were described and discussed in Part 1 and should by now be very familiar. The purpose of the fable of the three little phonemes was to show both how they are produced and where that production occurs in the end-to-end transmission of telephonic voice signals. The information disclosed in those narratives is summarized in Table 8-1,

TABLE 8-1

Determinants of Incidence and Severity of Perceptible Manifestations of Transmission Effects

Manifestation	Associated Performance Characteristics	Occurrence*
Low volume	Attenuation of speech power	B to C E to F F to G
Noise	Analog line noise	B to C F to G
Speech distortion	Quantizing noise	D to E IP1 to IP4′
	Amplitude clipping	D to D′
	Waveform deformation due to bit errors	D′ to E IP2 to IP4′
	Waveform deformation due to dropped frames	IP2 to E
	Phonemic clipping due to voice activity compression	D to D′ or IP1 to IP2
Echo	Echo path delay	B to J
	Echo path loss with network loss plan	B to C E to F H to I I to J
	Echo path loss with echo cancellation	E to H or E′ to H′
Excessive round-trip delay	Encoding time—PCM Encoding time—CELP Delay for packetization Jitter buffer delay Decoding time—CELP Decoding time—PCM	D to D′ IP1 to IP1′ IP1′ to IP2 IP3 to IP4 IP4 to IP4′ E′ to E
	Electrical transmission time—outbound	B to D D′ to E′ E to G
	Electrical transmission time—inbound	G to H H′ to I′ I to J
	Packet latency	IP2 to IP3

*Keyed to Fig. 8-2.

which displays the effects that determine the five manifestations identified at the beginning of this chapter, keyed to Fig. 8-2.

Multiple-Effects Models

Table 8-1, then, displays a compilation of various different performance characteristics that might affect the incidence and severity of degradations of quality experienced by users. Some models, such as the speech distortion models described in Chap. 7, produce measures that reflect multiple effects (various different kinds of waveform distortions and noise). The contributions of each effect in this case, however, cannot be inferred directly from the model.

The purpose of *multiple-effects models,* as opposed to models that simply reflect multiple effects, is to show how user perception of quality varies with measures of the individual performance characteristics whose combined effects may be encountered. Such models are invariably based on transforms of measures of performance characteristics into indicators that exhibit, or are expected to exhibit, a high degree of correlation with results of subjective user tests. When such indicators are produced, the model then supports prediction of likely user perception of quality as a function of the values of the different performance measures, thereby enabling analysis of how various different factors may be controlled to improve user perception of quality.

The following sections present examples of such models, that should represent, or at least typify, the models and techniques that are currently available for use in analysis of tradeoffs among performance characteristics or identification of specific causes of problems. As always, the objective here is to describe generally how such models work. Details of implementation are proprietary and, in many cases, represent intellectual property protected by patents.

Loss/Noise Grade of Service Model

One of the earliest multiple-effects models was the classical loss/noise grade of service model developed at AT&T Bell Labs. This model was developed from tests and analyses conducted during 1965 to 1970, and first reported internally at Bell Labs in 1976 in Ref. 5. It was later released for public use in the IEEE Global Telecommunications Conference of 1983 (Ref. 6).

It is a particularly good starting place for a description of multiple-effects models, not so much because it is still viable, but because

- It provides an easily understood illustration of the difference between a model based on a measure reflecting multiple effects, and a multiple-effects model in the sense in which the term is used here.
- It was the first model to employ *transmission ratings* as a device for handling the inevitable differences in subjective test results due to variations in test subjects, test environment, administration, etc.

Background

The principal motivation for the development of the loss/noise grade of service model was the inadequacy of the measure reflecting the combined effects of loss and noise in use at the time. The measure was the signal-to-noise ratio (SNR), defined as the ratio, expressed in decibels, of signal power to a commensurate measure of the power of noise. It had been used by telecommunications engineers for years and shown to correlate with user perception of quality well enough to serve as a reliable basis for deciding the acceptability of an end-to-end telephone service.

What was lacking with the SNR, however, was that it did not provide a basis for examining tradeoffs between the two performance characteristics whose joint effects it measured. Like an availability number without the associated mean time between failures and mean time to repair, the SNR could be interpreted to decide acceptability of service, but did not reveal anything that would support analysis of how to correct unacceptable quality.

Transmission Ratings

To develop a multiple-effects model that would correct this deficiency, the scientists at Bell Labs gathered together the results of three subjective user tests that had been conducted under controlled conditions. The three tests, identified by the location at which they were conducted, were Murray Hill (MH), and Homedale 1 and 2 (HO1, HO2). Each had been conducted under controlled conditions, to produce samples of opinion scores for conditions under which the connection loss and noise were known.

The plan of the study was to fit this data under some statistical assumptions to produce and estimate the variation in user opinion scores with loss and noise. This seemed simple enough. The envisioned data-fitting approach was workable but ambitious, since there was no such thing as a high-speed personal computer at the time.

However, a funny thing happened on the way to the envisioned model. When the fitting routines were applied independently to the three sets of test results, the fits for samples HO1 and HO2 were about the same, but substantially different from the fit for the MH sample. The differences and similarities passed statistical tests for significance, leading the analysts to conclude that "the subjects' ratings tended to be more critical in the two HO tests compared with the subjects in the MH test." (Ref. 6) This posed what should by now be the familiar conundrum of how to describe and gauge mean opinion scores, when the scores from any test can only be interpreted in light of test procedures, test conditions, and perspectives of the particular test subjects involved.

To get around the problem of such dependencies, the analysts at Bell Labs introduced the notion of a *transmission rating*. The idea was to define the transmission rating to be an indicator that would vary only with the loss/noise characteristics, but could nonetheless be shown to exhibit a high *correlation* with opinion scores produced from different subjective tests. Given such an indicator, it would then be possible to accept that subjective test results might vary with factors such as the test subjects, type of test, and conditions under which testing was conducted, and still derive meaningful interpretations from the different tests. Thus, for example, with a transmission rating, it would make no difference that one subjective test yielded a MOS of 3.5 for connections with 13 dB of loss and 30 dBrnC of noise while another yielded a MOS of 4.7. As long as the MOS values for all the other combinations of loss and noise tested exhibited a high correlation with the transmission rating, then it would be possible to interpret the transmission rating for a particular test by examining differences in MOS values and interpreting them in light of the particular perspectives of the test subjects, the test environment, etc. Moreover, given a high correlation between the transmission ratings and the predicted opinion scores from two different tests, it would be possible to transform the scores from one into scores for the other, and vice versa, in order to make commensurate comparisons for either test environment. Such relativity is anathema to those who want the MOS to be uniformly and absolutely interpretable, but the discussions of the pitfalls of MOS in Chap. 5 show that it is pretty much inescapable.

The indicator produced by Bell Labs was the loss/noise transmission rating, denoted R_{LN}. It was created by "anchoring" the indicator at two points that were to be used as boundary conditions for fitting other data. The points originally chosen are shown in Table 8-2, together with those eventually adopted as the standard by 1983.

In this scheme of things, any subjective test with mean opinion scores calculated for the 1976 anchor conditions, $\mu[15, 25]$, $\mu[30, 40]$ would, then, be fit to the transmission rating by setting

$$80 = R_{LN}[15, 25] = a + b\mu[15, 25] \qquad \text{and}$$
$$40 = R_{LN}[30, 40] = a + b\mu[30, 40]$$

and solving for a and b.

Application of these definitions and fitting the HO test data to capture variations in loss and noise, then, produced the classical loss/noise transmission rating scale:

$$R_{LN} = 147.76 - 2.257[(L_e - 7.2)^2 + 1]^{0.5} - 2.009N_F + 0.02037(L_e)(N_F) \qquad [8\text{-}1]$$

where L_e is loudness loss expressed in positive decibels, and N_F is the circuit noise N in dBrnC, increased by the power addition of 27.37 dBrnC.

Lessons Learned

Note that the inclusion of Eq. [8-1] here violates the avowed intent to avoid the specifics of models and definition of measures. The reason that

TABLE 8-2

Anchor Points for the Loss/Noise Transmission Rating

Date	Loudness Loss, dB	Circuit Noise, dBrnC	Transmission Rating
1976	15	25	80
	30	40	40
1983	10	25	80
	25	40	40

it is included here, however, is that it illustrates a characteristic found in many multiple-effects models for user perception of quality: The model in general, and the formulas for calculating relevant measures, in particular, are *intuitively obscure.*

Equation [8-1], for example, provides us with a painfully precise formula for the calculation of the loss/noise transmission rating that also happens to be completely arcane. The equation does suggest that R_{LN} decreases linearly as loss/noise increase, but the decrease is ameliorated by an amount that gets bigger as both the loudness loss and circuit noise increase. However, there is absolutely nothing in the structure of that formula that conveys an intuitive notion of how the combination of loudness loss and circuit noise shapes user perception of voice quality. There is no "natural" content in an indicator whose best value is 147.76, rather than some simple value like 100, which might suggest a grading scale. Moreover, because Eq. [8-1] can take on negative values, it is not clear how one might scale values of R_{LN} to produce such an indicator. Even to the mathematically sophisticated, there is no readily apparent reason for the transform of the L_e values. And, because of that transform and presence of the cross term, $(L_e)(N_F)$, there is no way to infer from the values of the coefficients the users' relative sensitivity to loss and noise.

All this means that the only way to validate Eq. [8-1] is to understand and accept as reasonable:

- Subjective test data on which it is based
- Assumptions as to the underlying distributions of values made to support the statistical fit of the data
- Underlying assumptions that enabled solutions of the mathematical equations
- Accuracy and reliability of the way the data were processed to produce the fits

R_{LN} is otherwise nothing more than an indicator whose semantic content is set by reference to the two anchor points, requiring us to know that $L_e = 15$ and $N = 25$ is "typical of a short intertoll connection" (Ref. 6), while $L_e = 30$ and $N = 40$ "represents an extreme condition of loss and noise that should rarely occur, even on long intertoll connections between long loops" (Ref. 6) in the United States long-distance telephone network in place in the 1965 to 1975 time frame.

The experience with the loss/noise grade of service model, then, suggests a number of inescapable characteristics of multiple-effects models of user perception of voice quality, described as follows:

Dependence on a Specific Set of Subjective Tests. The transmission rating R_{LN} and others of its ilk ultimately reflect the knowledge derived from a study of some particular set of subjective tests. In the case of R_{LN} the specific tests were MH, HO1, and HO2, and the model development was based on results from all three to produce Eq. [8-1]. There is, then, nothing to say that a different set of subjective tests might not have resulted in an entirely different formula. By the device of selecting the anchor points, the developers of this particular model, recognized and side-stepped the issue of dependency, by factoring out the subjective ratings altogether. Instead, the model as it is presented invites the user to specify the mean opinion score for the anchor points to calculate the values of R_{LN} that would apply for that environment. This means that the model cannot be used to predict mean opinion scores for particular conditions with respect to loss and noise without reference to a specific subjective test. This serves to amplify the pitfalls in interpreting the MOS described in Chap. 5.

Validation by Correlation. The dependency of predicted mean opinion scores on the particular sets of subjective test results demonstrates, in turn, that viable multiple-effects models will, in general, have two components:

- An *indicator* whose values show the combined effect of the factors considered, independent of the user perception of that effect
- A *transform* based on some subjective test data that predicts how users will assess voice quality as a function of that indicator

Now, the only transform that can be expected to replicate the original values of the subjective test is one derived from the same data as the indicator. We cannot, therefore, reasonably expect the predictions from the model to match the results from other subjective tests aimed at determining the users' response to the indicator, unless there are a number of stringent conditions satisfied, such as implementation of the same test protocol, use of a large number of test subjects, and precise replication of the conditions under which the original test was conducted. In practical terms, this means that *the measure of validity of a multiple-effects model is the* correlation *between the variations in the individual effects and observed measures of user perception of quality,* not the absolute difference.

This inescapable criterion for validity was recognized and built into the definition of R_{LN}. It is frequently overlooked in the never-ending search for a multiple-effects model that will accurately predict *the* MOS.

The E-Model

What has been described here as the loss/noise grade of service model was, in fact, but the first step in an overall effort reported in Ref. 5 that resulted also in formulas for producing R_E, the transmission rating factor for echo, and R_{LNE}, the transmission rating formula for the combined effects of loss, noise, and echo. The basic approach of defining transmission ratings for various different factors expected to affect user perception of voice quality and producing indicators of their combined effects has since been extended to numerous other factors.

The most recent extension of this kind at the time of writing of this book has resulted in the so-called E-Model, adopted by the ITU in Ref. 7 as an acceptable "computational model for use in transmission planning." In this model, the overall transmission rating R is defined by the equation

$$R = Ro - Is - Id - Ie - A \qquad \textbf{[8-2]}$$

where Ro = transmission rating based on the signal-to-noise ratio
Is = effects of a combination of impairments that "occur more or less simultaneously with the voice signal" (Ref. 7)
Id = effects of impairments due to delay
Ie = degradation of quality caused by low bit rate codecs
A = "compensation of impairment factors when there are other advantages of access to the users" (Ref. 7)

The extensions and modifications of the original transmission rating factor result in a transmission rating factor whose values now range between 0 and 100 and depend on the values of the many measures and indicators shown in Table 8-3.

The formulas for putting all the factors shown in Table 8-3 together to calculate the various transmission ratings and rating adjustments shown in Eq. [8-2] are clearly spelled out in Ref. 7. These instructions are accompanied by transforms that define the conversion of the indicator R into measures of *a* MOS and percentages of calls that are "good" or better (GOB) and "poor" or worse (POW).

However, if the descriptor *intuitively obscure* means "hard to grasp at the level of apprehending what is going on," the E-Model is intuitively opaque! Its validity must be accepted at face value on the basis of the 30 years of research that have gone into its development and its endorsement as an evolving standard by the ITU.

TABLE 8-3

Overview of Factors Whose Combined Effects Are Reflected in the E-Model Transmission Rating

Factor	Name	Subfactor	Reflects Effects of
Ro	Basic signal-to-noise ratio	SLR	End-to-end signal attenuation, expressed as a signal loudness rating
		No	Noise from a variety of sources including room noise, expressed as dBm using psophometric noise measurement
Is	Simultaneous impairment factor	*I*olr *I*st *Iq*	Low outbound volume Nonoptimum sidetone Quantizing distortion
Id	Delay impairment factor	*Id*,te *Id*,le *Id,d*	Talker echo Listener echo Excessive absolute delay, which can disrupt natural conversational rhythms
Ie	Equipment impairment factor	Type of codec	Speech distortion caused by low-bit-rate codecs, expressed as an assigned value for varieties of encoding collected in ITU G.113
A	Expectation factor	Type of connection	User accommodation of inferior quality in return for ability to use the telephone when: Moving about in buildings Moving about in a geographic area, or in a vehicle One end of the connection is in a hard-to-reach location Expressed as an assigned value to be taken from ITU G.113

E-Model Extensions to Packet-Switched Voice Services

While the E-Model has low inherent credibility, it does have two features that make it attractive as the starting point for producing user-perception models for packet-switched voice services. The first is that the E-Model reflects the combined effects of nearly all the factors shown in Table 8-3 to affect the incidence and severity of manifestations that shape user perception of quality. The contributions to perception of low volume are reflected in subfactors SLR and *I*olr. Noise effects are reflected in *No,* talker echo is reflected in *Id,*te, and the effects of excessive delay in conversational rhythms are reflected in *Id,d.* In addition, where there is information on it, the effects of the phonemic clipping that can occur are reflected in the *Ie* value assigned to codecs employing voice activity detection. In fact, comparisons of Tables 8-1 and 8-3 show that the only significant factor in a packet-switched voice service that is not somehow accounted for in the E-Model is the effect of dropped frames on user perception of the incidence and severity of speech distortion.

The other feature that makes the E-Model attractive is that it can easily be extended to include new effects as they are recognized. This feature is apparent in the cascade of subfactors exhibited in Table 8-3. Such ready extensibility means, in particular, that the model can be tailored to incorporate dropped frame rates as necessary to make it applicable to packet-switched voice services. The obvious alternative for doing this is to extend the model to include formulas expressing the *Ie* factors for each of the different codecs as a function of dropped frame rate. Another alternative would be to add to the model and Eq. [8-2] a new principal factor, say, *Ip,* for effects occurring only in packet-switched networks. This alternative is less intuitively satisfying, because it is reasonable to posit that the effects of dropped frames will vary with the type of codec, so that the effects would be an *Ie* subfactor. It is, however, viable within the overall structure of the E-Model.

Envisioned extensions of the E-Model to incorporate the effects unique to packet switching were still pending at the time of publication of this book, but may by now be promulgated in a more recent revision of ITU-T G.107 than the 12/98 version cited here (Ref. 7).

Voice Quality Evaluation System

Voice Quality Evaluation System (VQES) is the vanilla name given to a multiple-effects model developed by MCI/WorldCom for purposes of evaluating voice services that might involve packet switching as well as, or instead of, circuit switching. To enable exploitation of more than 10 years of testing with the service attribute test, this system was designed from the outset to be compatible with the subjective test protocol described in Chap. 6.

As suggested by the word *system* in the name, the VQES is a process supported by a combination of computer algorithms and formulas by which objective measurements of connection characteristics are transformed into estimates of subjective measures of quality. The overall process is illustrated in Fig. 8-4 and detailed here in terms of

- *Inputs*—the objective measures that might be used to characterize the end-to-end transmission effects in the service to be evaluated
- *Transforms*—the step-by-step processes by which a set of inputs is converted into estimates of measures that reflect likely user perception of quality and usability

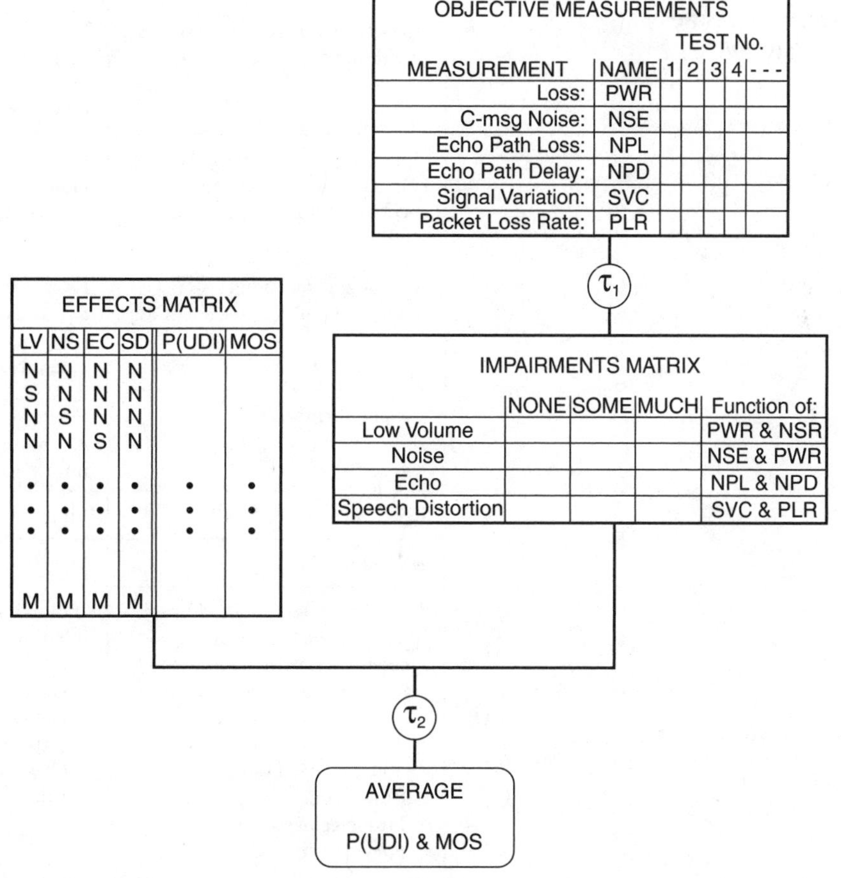

Figure 8-4 VQES translation process.

VQES Inputs

The basic data on which the VQES operates is a collection of objective measurements of end-to-end transmission effects that can be demonstrated from actual testing or reasonably assumed to characterize the service to be evaluated. The particular objective measurements required are ones known to affect the incidence and severity of the principal impairments described for SAT subjects or to directly affect the assessments of overall effect in accordance with the test protocol. The specific impairments and effects described for SAT subjects are summarized here in Table 8-4, together with examples of possible associated objective measures.

TABLE 8-4

Objective Measures Used in the VQES

Impairments	Principal Measure	Alternatives
Low volume	Speech power in dBm measured at the receiving handset	End-to-end signal attenuation in dB and statistics of speech power measured at a 0-dB test level point
Noise	C-message weighted noise in dBrnC measured at the receiving handset during a silent period	Psophometric weighted noise measured in dBm at the receiving handset
Speech distortion	Kurtosis of second differences of amplitudes of speech signals	PAMS or PESQ listening quality scores Intermediate psychoacoustic measures
	Dropped frame rates	Packet-by-packet jitter, to be used in conjunction with configuration information
Echo	Echo path loss	Singing point measurements
	Echo path delay	
Overall effect (U, D, I, N, or O)	Round-trip delay *added* by handling overheads and electrical transmission times not experienced on terrestrial circuit-switched connections in ms 500 ms per satellite hop Sum of one-way packet latencies across a packet-switched network	Individual measurements of inbound and outbound Electrical transmission time Encoding time Delay for packetization Jitter buffer delay Compression codec decoding time

The objective measures shown in Table 8-4 are characterized as *principal measures,* representing the objective measures for which the transforms are currently defined, or *alternatives.* The alternatives are common objective measures that might be used in lieu of the principal measures by resorting to models that establish relationships between the principal and alternatives, or by modification of the VQES to accommodate the alternative.

Note also in the table that some of the alternatives overlap two or three of the principal measure categories. This indicates that the alternative might replace more than one principal objective measure. For example, data on measures of the individual times might become a source for estimating both the echo path delay and the total round-trip delay. The similar overlap for PAMS, PESQ, and intermediate psychoacoustic measures indicates that these measures may reflect the effects of both noise and speech waveform distortions.

Transforms

Like the other multiple-effects models that have been described so far, the VQES utilizes two sets of transforms—one to translate objective measures of connection characteristics into indicators of subjective effects and another to translate the indicators into estimates of subjective measures of voice quality. Unlike the other models, however, the VQES does not combine effects to produce a single transmission rating. Instead, it preserves the association of the individual effects with the objective measures that affect them, to produce a multidimensional indicator comprising a frequency distribution for combinations of impairments represented in the input data set.

As shown in Fig. 8-4, the two-step process comprises

1. *Conversion to incidence and severity of impairments.* Transformation of inputs to produce sets of indicators for each input set
2. *Application of effects models.* Translation of the indicator sets into the measures of subjective assessment of voice quality derived from the SAT

The details of these are seen as follows.

Conversion to Incidence and Severity of Impairments The first step in the application of the VQES is production of the equivalent of transmission rating factors for the objective measurements in each of the sets of inputs. This is accomplished by defining a multiple-valued function for each measure of the form

$$\mathbf{P}_{\mathbf{Y}}[x] = P_N[x], P_S[x], P_M[x] \qquad \textbf{[8-3]}$$

where $\mathbf{Y}$ = particular objective measure

x = value of Y

$P_N[x], P_S[x], P_M[x]$ = estimated proportion of calls placed over a connection with objective measure $\mathbf{Y}$ having value x that SAT subjects would rate as having "none," "some," or "much," respectively, of associated impairment illustrated in Table 8-4

By definition, then, for any value of x, $P_N[x] + P_S[x] + P_M[x] = 1.0$, with each factor being greater than or equal to 0.

For example, if the principal objective measure were received speech power in dBm, shown in Table 8-4 to be associated with the low-volume impairment, then $P_N[-21]$, $P_S[-21]$, $P_M[-21]$ would be the estimated proportion of calls that would be rated, respectively, as having "none," "some," or "much" low volume in calls in which the received speech power was observed to be −21 dBm. Since a received speech signal level of −21 dBm is, if anything, much louder than normal, the expected values in this case would be ones with $P_N[-21]$ very close to 1.0, $P_S[-21]$ close to 0, and $P_M[-21] = 0$.

Many of the functions of the form shown in Eq. [8-3] have been derived from analysis of SAT conducted in conjunction with measurement of the characteristics of each test call. However, the functions of this kind can be developed completely independently from the SAT. The only question to be answered is, "Given a telephone call whose performance characteristic with respect to Y is measured at x, what is the probability that a person would rate the associated impairment as 'none' and what is the probability that a person would rate it as 'much'?"

Consequently, any data derived from subjective testing that will reliably answer this question for a number of different values of x can serve as a basis for the derivation of the functions. The only constraint is that there must be enough values of x for reasonable interpolation to define the P_Y function over the range of expected values of x.

The functions of the form displayed in Eq. [8-3] are then applied to convert inputs comprising sets of values of objective measures into estimates of the distribution of possible combinations of reports of the incidence and severity of impairments that might be reported in a SAT. The objective at this stage is to estimate the likely proportions of calls that would be assigned each possible combination in a large SAT for a set of values of objective measures:

$$\text{IP} = \{(V_j, N_j, D_j, \text{E1}_j, \text{E2}_j) \mid j = 1, \ldots, n\}$$

where the measures *V, N, D,* E1, and E2 are understood to be values or sets of values (e.g., E1 represents echo path loss and E2 represents echo path delay) of the objective measures assumed in associations like those shown in Tables 8-4 to affect the incidence and severity of reporting of low volume, noise, speech distortion, and echo in a SAT. The calculations are cumbersome, but straightforward. They proceed as follows:

1. As suggested in Table 8-5 which shows the 27 cases for low volume = "none," there are 81 possible combinations of SAT reports, comprising "none," "some," or "much" for each of the four impairments described for test subjects. These can be determined by application of the pattern shown in Table 8-5 and unambiguously identified by a series of four letters.
2. For each of the *n* sets of inputs, the required calculations are to apply the appropriate function to each of the objective values in the set to produce the associated P_N, P_S, P_M estimates, and then to multiply the individual results as illustrated in Table 8-6 to produce the proportions. As shown in Table 8-6, the necessary products parallel the pattern in Table 8-5 and can be identified from the four-letter case designations.
3. Finally the individual results from each of the *n* sets of values in the set IP are averaged, by adding up the various results and dividing by *n*. This means, for example, that P_{NNN}, the overall proportion of reports of type NNNN from the set IP, would be given by

$$\text{P}_{\text{NNNN}} = (1/n) \sum_{j=1}^{n} P_N[V_j] \cdot P_N[N_j] \cdot P_N[D_j] \cdot P_N[\text{E1}_j, \text{E2}_j] \qquad \textbf{[8-4]}$$

Application of Effects Models The transformation of the inputs into estimated proportions of each of the 81 combinations of "none," "some," and "much" responses then provides data in the form used in the T-SAT model described in Chap. 6. As described in Chap. 6, this model provides very large sample estimates of how users rate the quality and usability of a voice service as a function of users' descriptions of the impairments

TABLE 8-5

Pattern of Generation of Possible Combinations of Reports of Incidence and Severity

(Case of Low Volume = "None")

Low Volume	Noise	Speech Distortion	Echo	Case Designation
			N	NNNN
		N	S	NNNS
			M	NNNM
			N	NNSN
	N	S	S	NNSS
			M	NNSM
			N	NNMN
		M	S	NNMS
			M	NNMM
			N	NSNM
		N	S	NSNS
			M	NSNM
			N	NSSN
N	S	S	S	NSSS
			M	NSSM
			N	NSMM
		M	S	NSMS
			M	NSMM
			N	NMNN
		N	S	NMNS
			M	NMNM
			N	NMSM
	M	S	S	NMSS
			M	NMSM
			N	NMMN
		M	S	NMMS
			M	NMMM

Note: N = "none," S = "some," M = "much."

noted. It shows for each of the case designations like those in Table 8-6 the expected value of the MOS, P[UDI], and PGoB derived from composite SAT data representing more than 150,000 calls in which the "none," "some," and "much" responses of each of the impairments were recorded together with the opinion score and the description of effects as "unusable," "difficult," "irritating," "noticeable," or "none."

The estimation of the overall MOS for an input set thus becomes a straightforward application of the T-SAT model obtained by weighting each of the possible cases with the proportions like those shown in Table 8-6. That is, if MOS(NNNN) denotes the T-SAT value of the MOS for the case of all "none" reports, and P_{NNNN} is the proportion defined in Eq. [8-4], then the weighted entry for that case designation is just the product: $MOS(NNNN) \cdot P_{NNNN}$. The overall MOS is then estimated by calculating products of this type for all 81 case designations, and adding them up.

The overall proportion of calls expected to be rated as "unusable," "difficult," or "irritating" associated with the impairments, $P_I[UDI]$, is similarly calculated by weighting the cases by their proportions to get an average for the input set. The $P_I[UDI]$ is then increased to reflect the expected effect of the round-trip delay. This is accomplished by using a model derived from the satellite era that predicts the proportion of the population that will notice or experience disruptions of conversational rhythms due to excessive round-trip delay as a function of the delay. Denoting this model value by $P_D[UDI \mid \delta]$, where δ denotes a round-trip delay, the overall P[UDI] for an input set for which the round-trip delay was δ is then estimated by setting

$$P[UDI] = P_I[UDI] + (1 - P_I[UDI]) \cdot P_D[UDI \mid \delta] \qquad \textbf{[8-5]}$$

Equation [8-5] is an example of a probability model for estimating the combined effect of two independent contributing causes. In this case, the model is based on the idea that the round-trip delay will be a contributing factor that will increase the overall proportion of calls rated "unusable" (U), "difficult" (D), or "irritating" (I), by adding calls that would not be rated U, D, or I due to other impairments, but would be so rated because of the round-trip delay.

Another probability model of this kind that may be applied within the structure of the VQES works like this: Suppose that there are two independent factors *A* and *B* that contribute to the perceived incidence and severity of the same impairment defined for SAT subjects, and let $P_X[A]$ and $P_X[B]$, where X = N, S, M, denote the conversions of either. Then $P_X[A\&B]$, the proportions expected to be in each rating category as a result of the combined effect, can be estimated by the following equations:

TABLE 8-6

Calculation of Expected Proportions of Occurrence of Combinations of Reports of Incidence and Severity of Impairments in a Service Attribute Test for Input Set (V_j, N_j, D_j, $E1_j$, $E2_j$)

(Case of Low Volume = "None")

Low Volume V_j	Noise N_j	Speech Distortion D_j	Echo Factors $E1_j$, $E2_j$	Proportion	Case Designation
			$P_N[E1_j, E2_j]$	$P_N[V_j] \cdot P_N[N_j] \cdot P_N[D_j] \cdot P_N[E1_j, E2_j]$	NNNN
		$P_N[D_j]$	$P_S[E1_j, E2_j]$	$P_N[V_j] \cdot P_N[N_j] \cdot P_N[D_j] \cdot P_S[E1_j, E2_j]$	NNNS
			$P_M[E1_j, E2_j]$	$P_N[V_j] \cdot P_N[N_j] \cdot P_N[D_j] \cdot P_M[E1_j, E2_j]$	NNNM
			$P_N[E1_j, E2_j]$	$P_N[V_j] \cdot P_N[N_j] \cdot P_S[D_j] \cdot P_N[E1_j, E2_j]$	NNSN
	$P_N[N_j]$	$P_S[D_j]$	$P_S[E1_j, E2_j]$	$P_N[V_j] \cdot P_N[N_j] \cdot P_S[D_j] \cdot P_S[E1_j, E2_j]$	NNSS
			$P_M[E1_j, E2_j]$	$P_N[V_j] \cdot P_N[N_j] \cdot P_S[D_j] \cdot P_M[E1_j, E2_j]$	NNSM
			$P_N[E1_j, E2_j]$	$P_N[V_j] \cdot P_N[N_j] \cdot P_M[D_j] \cdot P_N[E1_j, E2_j]$	NNMN
		$P_M[D_J]$	$P_S[E1_j, E2_j]$	$P_N[V_j] \cdot P_N[N_j] \cdot P_M[D_j] \cdot P_S[E1_j, E2_j]$	NNMS
			$P_M[E1_j, E2_j]$	$P_N[V_j] \cdot P_N[N_j] \cdot P_M[D_j] \cdot P_M[E1_j, E2_j]$	NNMM
			$P_N[E1_j, E2_j]$	$P_N[V_j] \cdot P_S[N_j] \cdot P_N[D_j] \cdot P_M[E1_j, E2_j]$	NSNM
		$P_N[D_j]$	$P_S[E1_j, E2_j]$	$P_N[V_j] \cdot P_S[N_j] \cdot P_N[D_j] \cdot P_S[E1_j, E2_j]$	NSNS
			$P_M[E1_j, E2_j]$	$P_N[V_j] \cdot P_S[N_j] \cdot P_N[D_j] \cdot P_M[E1_j, E2_j]$	NSNM
			$P_N[E1_j, E2_j]$	$P_N[V_j] \cdot P_S[N_j] \cdot P_S[D_j] \cdot P_N[E1_j, E2_j]$	NSSN
$P_N[V_j]$	$P_S[N_j]$	$P_S[D_j]$	$P_S[E1_j, E2_j]$	$P_N[V_j] \cdot P_S[N_j] \cdot P_S[D_j] \cdot P_S[E1_j, E2_j]$	NSSS
			$P_M[E1_j, E2_j]$	$P_N[V_j] \cdot P_S[N_j] \cdot P_S[D_j] \cdot P_M[E1_j, E2_j]$	NSSM

			$P_N[E1_j, E2_j]$	$P_N[V_j] \cdot P_S[N_j] \cdot P_M[D_j] \cdot P_M[E1_j, E2_j]$	NSMM
		$P_M[D_j]$	$P_S[E1_j, E2_j]$	$P_N[V_j] \cdot P_S[N_j] \cdot P_M[D_j] \cdot P_S[E1_j, E2_j]$	NSMS
			$P_M[E1_j, E2_j]$	$P_N[V_j] \cdot P_S[N_j] \cdot P_M[D_j] \cdot P_M[E1_j, E2_j]$	NSMM
			$P_N[E1_j, E2_j]$	$P_N[V_j] \cdot P_S[N_j] \cdot P_M[D_j] \cdot P_M[E1_j, E2_j]$	NMNN
		$P_N[D_j]$	$P_S[E1_j, E2_j]$	$P_N[V_j] \cdot P_M[N_j] \cdot P_N[D_j] \cdot P_S[E1_j, E2_j]$	NMNS
			$P_M[E1_j, E2_j]$	$P_N[V_j] \cdot P_M[N_j] \cdot P_N[D_j] \cdot P_M[E1_j, E2_j]$	NMNM
			$P_N[E1_j, E2_j]$	$P_N[V_j] \cdot P_M[N_j] \cdot P_S[D_j] \cdot P_M[E1_j, E2_j]$	NMSM
	$P_M[N_j]$	$P_S[D_j]$	$P_S[E1_j, E2_j]$	$P_N[V_j] \cdot P_M[N_j] \cdot P_S[D_j] \cdot P_S[E1_j, E2_j]$	NMSS
			$P_M[E1_j, E2_j]$	$P_N[V_j] \cdot P_M[N_j] \cdot P_S[D_j] \cdot P_M[E1_j, E2_j]$	NMSM
			$P_N[E1_j, E2_j]$	$P_N[V_j] \cdot P_M[N_j] \cdot P_M[D_j] \cdot P_N[E1_j, E2_j]$	NMMN
		$P_M[D_j]$	$P_S[E1_j, E2_j]$	$P_N[V_j] \cdot P_M[N_j] \cdot P_M[D_j] \cdot P_S[E1_j, E2_j]$	NMMS
			$P_M[E1_j, E2_j]$	$P_N[V_j] \cdot P_M[N_j] \cdot P_M[D_j] \cdot P_M[E1_j, E2_j]$	NMMM

N = "none," S = "some," M = "much."

$$P_N[A\&B] = P_N[A] - P_N[A] \cdot (P_S[B] + P_M[B]) \quad \textbf{[8-6a]}$$

$$P_S[A\&B] = P_S[A] + P_N[A] \cdot P_S[B] - P_S[A] \cdot (\alpha \cdot P_S[B] + P_M[B]) \quad \textbf{[8-6b]}$$

$$P_M[A\&B] = P_M[A] + P_N[A] \cdot P_M[B] + P_S[A] \cdot (\alpha \cdot P_S[B] + P_M[B]) \quad \textbf{[8-6c]}$$

where α is a factor representing the proportion of calls for which the presence of both factors *A* and *B* at levels that would result in a "some" response were they present alone would result in a "much" response for the associated impairment. For precision, the value of α can be determined empirically. However, it can also simply be assumed to equal ½ without greatly affecting the VQES model predictions.

Limitation

As might have been recognized by those steeped in the arcane arts of measurement and evaluation of voice quality, the VQES does suffer from one inherent limitation due to its structure. The problem is that functions of the form shown in Eq. [8-3] cannot result in anything worse than $P_N = P_S = 0$ and $P_M = 1.00$. This means that the lowest MOS value that can be assigned to a call exhibiting poor values of only one of the associated objective measures is the MOS for samples of calls in which that impairment was reported as "much." As a consequence, the VQES-predicted MOS and P[UDI] values cease to be interpretable when the threshold value for $P_N = P_S = 0$ and $P_M = 1.00$ is reached.

For example, suppose that analyses show that the proportion of reports of "much" noise grows to 1.0 at about 55 dBrnC. Then application of the VQES will predict the T-SAT MOS value for the case designator NMNN for all calls in which the only impairment was noise measured at 55 dBrnC or greater. If that MOS value were, say, 2.4, then the VQES will assign the same 2.4 to a call with, say, 75 dBrnC of noise.

This is an inescapable result of the way that objective measures are transformed to enable application of the effects models used by the VQES. It does not, however, imply that the model is invalid or unusable. It only means that, like many models of this kind, there is a range of input values for which meaningful changes are reflected, and input values outside of that range result in meaningless outputs.

Moreover, the inherent limitations in the range of inputs that will result in meaningful outputs do not degrade the utility of the model, because the meaningless outputs occur only for very bad conditions. For example, it makes little difference whether the P_M values become 1.0 because of 55 or 75 dBrnC of noise. That a service in which nearly all users rate every connection as having "much" noise is unacceptable will be demonstrated by such wide variance from MOS and P[UDI] values for an acceptable service, that there is little additional value in determining whether it is "bad," "rotten," or "terrible."

Features

As indicated earlier, the characteristic that makes the VQES particularly attractive for use by WorldCom is that it is based on the results of years of subjective testing with which the company is familiar and comfortable. This is not, however, the only reason that the VQES has endured as WorldCom's preferred multiple-effects model as the E-Model has evolved into the ITU standard. Rather, its continued use is due both to its origins and to the features described in Chap. 5—inherent credibility, extensibility, and manipulability—that greatly enhance the utility of any model. The realization of each of these features in the VQES is now described briefly, to illustrate these concepts and stress their importance in assessing the utility of any multiple-effects model that may come along after the ones described here.

Inherent Credibility Unlike the transmission factors from the classical multiple-effects models and their extensions in the E-Model, the indicators produced by the functions of the kind represented by Eq. [8-3] are ones that can be readily tested against experience. It is very hard, for example, to ascribe any concrete meaning to the assertion that the basic signal-to-noise ratio R_o, in the E-Model is given by the equation

$$R_o = 15 - 1.5\,(\text{SLR} + N_o) \qquad \textbf{[8-7]}$$

where SLR is a send loudness rating ascribed to the transmitted speech power measured at a 0 test level point, and N_o is the power addition of four different kinds of noise measurements, or to the information that the value of R_o for the best service is 94.2. In contrast, equations of the form of Eq. [8-3] enable a prospective user of the VQES to test the outputs against common measurements, to produce, for example, concrete implications like

When the end-to-end circuit noise is 20 dBrnC, nearly all users will report "none" for noise.

When the received speech power is −25 dBm, almost no one will complain of low volume, but when it drops to −48 dBm, 90 percent of telephone users will report that the severity of low volume is "much."

Such derivations from the VQES indicator functions for each attribute can be easily tested by persons conversant with telephony measurements to verify that the basis for the model is both reasonable and consistent with their experience. Such ready testability of the model indicators is a practical realization of the kind of natural ease of understanding and apprehension of the model functions that defines inherent credibility.

Extensibility Because the intermediate indicators in the VQES are tied to user perceptions of specific kinds of impairments, it is very easy to extend the model to reflect different independent effects. For example, addition of dropped frame rates as an effect that would be encountered in packet-switched networks was accomplished by

1. Observing that dropped frame rates would be an independent factor affecting user perception of speech distortion
2. Conducting subjective user tests to determine empirically how users are likely to rate the speech distortion for different known dropped frame rates
3. Analyzing the data from those subjective user tests to produce the vector-valued function $P_{\mathrm{DFR}}[x]$, which yields, for any value of the dropped frame rate x, the expected proportions of reports of "none," "some," and "much" speech distortion
4. Applying Eq. [8-6] to calculate the combined effects of other identified causes of speech distortion (in this case measured by kurtosis as defined in Table 8-4) and dropped frame rates

This suggests the ease with which the VQES can be extended to reflect new effects or to change the objective measures that define an input set. What is neither as obvious nor as easily explained (and will, therefore, not be done so here) is the fact that in many cases one can calibrate the VQES for any new environment without conducting additional subjective testing. For example, suppose the $P_N[x]$, $P_S[x]$, $P_M[x]$ functions for dropped frames currently in use were derived from subjective testing of a packet-switched voice service using a G.729 codec. Then, because of the structure of the VQES, those functions can be adjusted for a service using a G.711 codec without conducting further subjective tests. Instead, it is possible to analyze sets of objective measures required for the VQES collected from an environment using the G.711 codec to produce transforms that will define the $P_N[x]$, $P_S[x]$, $P_M[x]$ functions for that environment.

An analogous technique can be used to change the objective measure used for a particular aspect of the VQES. For example, suppose we wanted to produce a version of the VQES in which noise is measured by the psophometric noise weighting technique preferred for the E-Model instead of the C-message noise measurements currently used. Then, while it might be desirable to conduct subjective tests to produce the $P_N[x]$, $P_S[x]$, $P_M[x]$ functions for *P*-weighted noise, a viable alternative would be to analyze data sets comprising values of *P*-weighted noise, together with the other principal VQES measures shown in Table 8-4.

Manipulability Finally, it should be noted that all three of the examples of multiple-effects models described here are inherently manipulable, or else they could not be held up as examples of such models. The reason is that the principal purpose of any multiple-effects model is to afford its users the opportunity to examine tradeoffs among different effects, including those that have not yet occurred in operational services. In order to do this, it must be possible to

- Use some actual data to describe the existing service and then examine what would happen were various hypothetical combinations of effects prevalent in the service.
- Use the model to estimate and compare the relative impact of different contributing effects, acting alone or in concert on user perception of voice quality.

All three of the models described here support these applications. However, as of the time that this was written only the VQES fully supports the specific purpose of examining tradeoffs among effects that are possible in a packet-switched transport. Moreover, it does so in a particularly straightforward manner, by explicitly including provisions for objective measures of the incidence of dropped frames and increases in round-trip delays due to packet latency, which reflect the principal deleterious effects associated with packet switching.

This also means that the VQES encompasses all the machinery to readily characterize user perception of quality of voice services carried over the three service models—hybrid transport, packet-switched telephony, and interactive multimedia exchange—described in Chap. 2. For example, for hybrid transport it will be necessary to examine input sets comprising representative values for all the principal objective measures shown in Table 8-4. For packet-switched telephony, the same model can be used by defining input sets in which the values of the objective measures associated with connections across the PSTN are each set to assure predictions of P_N are very close to 1.0. And, for interactive multimedia exchange, the difference in user expectations can be readily accommodated as appropriate by changing the function that predicts the proportion of exchanges that will be rated as "unusable," "difficult," or "irritating," to reflect a greater tolerance for delayed responses.

Comment

Again, although it may not seem like it, the purpose of the description of the important features of the VQES here is not to posit its superiority. Rather, what should be gleaned from these discussions is what features like this really look like, so that their presence or absence in any multiple-effects model can be readily recognized and appreciated in the decision of which model to use for examination of tradeoffs. As will be seen in Chap. 9, such features also play a very important role in determining whether a particular model can be readily used for a particular application.

CHAPTER 9

Applications

The preceding chapters have suggested a variety of techniques that might be useful for estimating likely user perception of the quality of voice carried over packet-switched services. This chapter examines possible applications of these techniques.

More formally, the purpose of this chapter is to illustrate possible applications of measurement and evaluation of packet-switched voice services, by

- Characterizing the possible requirements for information on user perception of voice quality in addressing questions of design, management, and operation services using packet-switched transport
- Illustrating how the measures, models, and tests that have been described so far might be brought to bear in meeting those requirements

To accomplish this, the presentation begins with a description of a formal process recommended for use in any analytical effort of this kind and shows how that process is realized for the general problem of evaluation of packet-switched voice services. It then proceeds to detail the application of the process to a number of specific issues whose resolution requires information on likely user perception of voice quality.

Analytical Procedure

As envisioned here, the principal uses of the analysis techniques that have been examined so far will be to support decision making with one of two products:

- *Voice quality evaluation,* showing whether the quality of a voice in a particular service will be found to be acceptable to its users. This involves definition of measures on which the evaluation is to be based, quantification of those measures, and interpretation of the results to answer the basic question of whether the voice quality will be acceptable to the service users.
- *Service analysis,* showing what must be achieved in a voice service employing packet-switched transport to assure that the service will be acceptable to users and stay that way. This inevitably involves development of a means of expressing criteria for user satisfaction

in terms of measures of performance characteristics that can be controlled by service providers.

In casting about for the best technical approach for these activities, there is a temptation to turn the selection process into a choice of which technique shall become *the* method to be used for all evaluation and analysis. What should happen, however, is that the methodology for any voice quality evaluation or service analysis should be deferred until five basic questions have been answered.

1. *What is the objective?* Answered, for example, by characterizing the kind of decision making to be supported, the needs of the audience that must accept and act on the information, and the implied constraints on the basis and nature of the information that must be produced to satisfy the needs of the audience.

2. *What must be reflected in the measures?* Answered by specifying the service attributes that must be examined in order to produce answers to the specific questions expected from decision makers, and determining from that specification which measures of quality must be quantified.

3. *What are the alternatives for quantifying the selected measures?* Answered by identifying calculations, including applications of models, that might be involved in determining meaningful values of the measures selected for the effort and developing a list of data elements required for those calculations. This step is crucial to the cost-effectiveness of the effort, because the resultant specification of data elements will determine the utility of data sources having vastly different data-acquisition costs.

4. *What are the viable options for acquiring data that might be used to quantify the selected measures?* Answered by enumerating possible sources for each of the data elements defined in answering question 3, together with the data-acquisition methods and, where possible, any means for meaningfully combining data elements acquired from different sources.

5. *What is the most cost-effective combination of data acquisition, handling, processing, and analysis to produce the required information?* Answered, for example, by describing the data-collection and analysis plan that appears to make the best use of readily available data, poses the minimal data-acquisition effort, and minimizes calculation time when speed of quantification of measures is an important consideration or data storage requirements when the size of the databases will be a constraint.

To illustrate this process and establish a basis for its application, this section provides exemplary answers to the first four of these general questions for the specific case of voice quality evaluation and service analysis for packet-switched voice services. The next section then illustrates the results of the process by describing the most cost-effective alternatives for a variety of specific examples taken from applications to VoIP.

Objectives

For purposes of illustrating how consideration of the objectives of an analysis of voice quality may shape the selection of measurement techniques, the following subsections examine the objectives and implications of five distinctly different functions:

- *Network design.* Determination of what performance is required in a packet-switched transport to assure that the voice quality will be acceptable.
- *Market assessment.* Determination of user expectations of, and customer requirements for, packet-switched voice services.
- *Service characterization.* Description of the expected performance and associated user perception of quality of a packet-switched voice service.
- *Service validation.* Verification that a packet-switched voice service as installed and configured for turnover to a customer conforms to service provider standards.
- *Quality monitoring.* Continuous evaluation of the quality of voice signals transported via a packet-switched voice to detect possible deterioration of quality.

In what follows, each of these functions is described and analyzed to show how such functional disparities may dictate different voice quality measurement and service analysis objectives.

Network Design The objective of network design efforts is to determine how to configure packet-switched transport to meet voice quality requirements with the least cost in terms of investments in telecommunications capacity and call handling technology. The principal questions to be resolved are

- What codec and codec configuration should be used for voice encoding and decoding?

- Which packetization scheme should be employed for voice handling?
- What QoS techniques and levels should be used for assuring that voice packets receive adequate preferential handling and protection?
- How much capacity on customer accesses and egresses is needed to assure acceptable handling of the expected traffic?

The users of such information will be those responsible for designing and engineering the service. Consequently, the answers to such questions must be couched in terms of system features, functions, and performance characteristics affecting voice quality that can be controlled by design, configuration, and implementation. Moreover, to be most useful, the answers will be expressed in terms of *criteria* or standards to be applied to each element. This means, for example, that features and functions will be best described by descriptions of acceptable alternatives for transmission speeds and equipment, choice and configuration of codecs, etc. Specifications of performance characteristics will similarly be most useful when they are expressed in terms of fixed thresholds for a collection of measures, determined in such a way as to assure acceptable voice quality when none of the measures violates its threshold. The most workable expression of performance requirements will thus be a set of statements like

- Expected noise should not exceed x dBrnC.
- End-to-end loss should be no less than y dB and no greater than z dB.
- Dropped packet rate at the far side of the codec used for packet-switched transport should be no greater than d.

The objective is to arrive at values of x, y, z, d, etc., that can be readily achieved through element-by-element control of performance characteristics in the system design and architecture. Specification of thresholds for each performance characteristic like this is very conservative, because there will be combinations of values in which some thresholds are violated, but the voice quality is still acceptable. The criteria do, however, serve as a convenient partitioning of effects to create a design budget for the various different factors that might degrade voice quality. Such an allocation of degradation due to various different contributing factors is much more useful than, say, simply trying to design the system to achieve a particular transmission rating.

The development of performance criteria will require an ability to estimate voice quality for a variety of equipment options and configurations supporting a wide range of different performance expectations. This means, almost by definition, that the network design must be supported by a readily extensible multiple-effects model, capable of mixing empirical results with presumed values of performance characteristics derived from analysis of technical characteristics and specifications.

Moreover, since the actual performance characteristics will often be associated with equipment or configurations that have never been tested in an operational environment, it will be imperative that *none of the indicators that translate objective measurements in the model is unchangeably based on a particular type of connection, test condition, transmission medium, or voice encoding scheme.* Otherwise, the examination of alternatives will be masked by the assumed basis for the assignment of the indicator value.

Market Assessment When they are initially deployed, packet-switched voice services will be new and unfamiliar. Service providers will nonetheless have to be able to advertise, represent, sell, and support those services as if they were a century old. This means that marketing personnel will have to formulate their sales strategies by adopting the perspective of prospective customers, anticipating their desires and concerns, and acting out the customers' decision-making process for choosing from among competing service offerings.

To accomplish this end, marketing personnel must have a very good idea of

- The features, functions, performance, and quality that customers value in their telecommunications services
- How the packet-switched voice service to be offered stacks up against these characteristics
- The comparative economies of the newly offered service relative to the systems currently providing comparable services

Against these very wide ranging needs for information to support marketing and sales, results of voice quality evaluation can provide the basis for answering questions in three critical areas.

- *Quality expectations.* How does the voice quality expected in our new packet-switched service compare with existing services? Which impairments will be reduced or avoided? Which will be more frequent or severe? What is the likelihood that the quality of voice over the packet-switched network will be noticeably different from conventional voice services?

- *Competitive posture.* How does it compare with packet-switched voice services being offered by the competition?
- *Impact.* How does achievement of acceptable voice quality affect capacity and throughput requirements? What are the cost and quality tradeoffs?

Because the answers to these questions must resonate with the customers raising them, marketing must be positioned to answer them in terms commensurate with the customer's decision making. This means that the marketing assessment function must be equipped with comparative measurements of voice quality that run the gamut from the associated factors employed for network design through those used for service characterization described in the next section. For example, when a customer employs the PESQ measurement for voice quality monitoring, the answers to the questions previously stated should be based on a comparison of PESQ values. When a particular customer's chief complaint is occasional low volume on calls, the marketer should be able to show comparisons of the expected incidence and severity of low volume among the existing service, the offered packet-switched service, and the packet-switched service being offered by the competition.

In addition, to ensure credibility, the data for making such comparisons should be drawn from tests in an environment that closely resembles that in which the customer will use the service. Failing this, the comparisons should, where possible, be based on models that allow for adjustments that can be made to closely approximate that operating environment. In practical terms this means that market assessment must ultimately be supported with a good model for gauging voice quality as a function of access and termination capacity and traffic loading.

Service Characterization The purpose of service characterization is to create an accurate description of what users can expect to experience in using that service. The principal audience for such descriptions will be the community of prospective or actual users and customers seeking reassurances that a packet-switched voice service will achieve and maintain acceptable voice quality. The descriptions achieved must be

- *Clear and precise,* allowing for little possibility of misinterpretation
- *Meaningful,* portraying the expected quality of the voice service in terms that readily convey to users what they can expect to hear and experience

- *Inherently credible,* so that users without technical expertise can feel comfortable with the way that the information was produced
- *Scientifically defensible,* because the representations of voice quality in the service characterization will be communicated in representations by marketing personnel and advertising that may be publicly challenged by the competition

Taken together these criteria suggest that the preferred basis for service characterization will be results of tests of calling placed through an operational, realistically loaded service of the type characterized. The testing would ideally be subjective, but a scientifically defensible multiple-effects model applied to objective measures might be an acceptable alternative, as long as the model has high inherent credibility.

Because the audience for service characterization is largely composed of persons who do not understand, and could care less about technology, the descriptions of voice quality derived from the data collected should be as nontechnical as possible. Some may be comfortable with descriptions of expected MOS and measures of usability, as long as those values are clearly anchored to similar values for some familiar service. However, the preferable approach would be to couch the service characterization in terms of the *interpretation* of tests of voice quality, describing the expected differences in *qualitative,* rather than quantitative, terms. The quantitative results might not be left aside altogether, but they should be used only to elaborate or exemplify the comparisons claimed.

All this means, for example, that instead of reading:

> The packet-switched voice service *X* can be expected to exhibit
>
> - Some or much speech distortion in 5 percent of the calls as compared to 3 percent currently experienced in the PSTN
> - Noticeable effects of excessive delay in 2.5 percent of the calls
>
> but will have noticeable noise or volume problems in 18.5 percent fewer calls. The result is a MOS of 3.48 as compared with 3.54 for the PSTN.

the characterization of this hypothetical service should read:

> The packet-switched voice service *X* will exhibit slightly higher incidences of speech distortion and disruption of conversations due to longer than normal delays between what you say and the response from the distant party. The greater than normal delay will be unusual enough to be noticed as a difference in the packet-switched service, but should not occur often enough to represent a problem. Moreover, the differences in speech distortion will not generally be noticeable, while the improvement in reduction

of noise and volume problems will be. Given these conditions, a user may notice slight differences between the packet-switched voice service and the familiar PSTN service, but should not find those differences great enough to cause dissatisfaction with voice quality.

Service Validation The function of service validation is to verify that services turned up are free of obvious defects that would prevent attainment of the best voice quality achievable within the constraints of the traffic loads and allocated capacity. The principal users of the information are network design and architecture agents responsible for provisioning the transport network and for sizing and installing customer accesses.

The major question to be answered from service validation is whether newly installed node-to-node connections in the packet-switched network are functioning to full potential. To avoid unnecessary delay in turning up the connections for use, the tests and analysis involved in doing this should be something that can be completed in a matter of a few days. Because there may be many tests of this kind required as use of the service expands, service validation requires the support of

- Easily implemented, very inexpensive data-acquisition capabilities
- Very robust tools for analyzing data to infer the condition of newly activated routes

This implies, in particular, that the raw data acquired must, at the very least, enable determination of packet loss, dynamics of jitter, and expected round-trip delay across a sample of origin-to-destination relays that traverse the newly activated route.

The critical support for service validation is thus the capability to acquire the required data over any selected origin-to-destination route, rather than timeliness or ready interpretability of the data captured. The basis for validation may, then, be something as involved as processing the data to produce inputs for a multiple-effects model, or as simple as producing a quickly calculated indicator used for quality monitoring. Either way, it will be essential also to preserve *raw* test data in a readily accessible form to support continuing analysis aimed at enhancing the effectiveness of the validation procedure by refining measures, quantifiers, and criteria for acceptance.

Quality Monitoring The principal functions of quality monitoring are to develop timely indications of deterioration of voice quality, together with information that might support corrective actions, and to reassure customers that voice quality is being effectively maintained, by routinely reporting results. Effective quality monitoring is particularly important in the case of packet-switched voice services, because quality is much more volatile than it is in conventional services. In a circuit-switched voice service the deterioration of voice quality is invariably attributable to the malfunction or misalignment of one of perhaps thousands of components that might be used to establish the connection. When such a problem emerges, it persists or worsens until it is detected and corrected, producing only sporadic instances of poor quality over a particular set of origin-to-destination connections. A packet-switched service will experience similar isolated problems that must be detected and corrected. However, the preponderance of quality problems will occur as a result of conditions in the network that can affect the quality of all calls active at the time. For example, because of the way packets are delivered in a packet-switched network, a malfunctioning router that is dropping or erroneously queuing voice packets can affect the quality of all calls handled through that router. Similarly, transient congestion at the origin or destination of the kind that would produce fast busy signals in a circuit-switched network will affect the voice quality on *all* calls into or out of the affected edge router.

The principal users of quality monitoring information will be operations personnel tasked with maintaining service quality. To support their needs the data acquisition and processing for quality monitoring must be robust, ubiquitous, and highly efficient, capable of generating near-real-time values for a large number of different possibilities for call origin and destinations. The expected day-to-day operational uses of a quality monitoring system therefore create requirements for

- Capabilities to acquire and process data for a large number of different origin-to-destination possibilities
- Timely notification of indications of degraded voice quality
- High reliability of those indications, so that operations personnel do not waste time pursuing false alarms

This implies that the basic quantifiers produced in a voice quality monitor system need not, and probably should not, be fine-grained estimates of user perception of voice quality. Rather, for purposes of computational efficiency, what is produced should be a reliable *indicator* of quality that can be calculated and interpreted quickly enough to generate timely alarms.

The system will, moreover, be most useful when the data that are acquired and processed to produce alarms of possible deterioration of voice quality can be

- Readily interpreted to support identification of likely causes of, and corrections for, the conditions generating the alarm
- Efficiently stored and retrieved to support analysis for trends or more subtle indications
- Easily massaged to satisfy reporting requirements for measuring and reporting voice quality specified in service-level agreements

This, together with requirements to monitor a large number of different origin-to-destination call possibilities implies that the data to be analyzed should be captured in very compact data structures that can be easily manipulated to aggregate data sets, produce estimates of subjective measures of voice quality, etc.

Measures

Having determined the objectives, the next step in strategizing a voice quality evaluation or service analysis will be to ascertain what measures will best answer the specific questions that the decision makers are likely to raise. This is accomplished, in general, by examining the audience and its objectives to

- Identify the principal *concerns* of the cognizant decision makers
- Select or define generic measures that most directly relate to those concerns and, where necessary, most clearly and immediately relate to familiar measures of performance

For the case of quality of service, or some particular aspect of quality of service, the basic concern of the decision makers will be how well a particular service performs relative to expectations. The measures that most directly relate to this concern are dependent on *whose* expectations are being addressed, creating what can be conveniently thought of as three broad classes of measures of QoS identified in Ref. 1.

- *Intrinsic.* Comprising measurements of quality of service relative to the expectations of decision makers who design and operate the systems that deliver the service
- *Perceived.* Comprising measures that quantify quality relative to the expectations of the persons who use the service

- *Assessed.* Comprising measures that reflect the expectations of persons who must consider the economics of the service and deal with the service provider on matters of billing, ordering, correction of problems, etc.

The generic measures of voice quality in each category are summarized in Table 9-1. With one exception, the measures that might be appropriate for packet-switched voice are self-explanatory or have been defined previously. The exception is *service stability* shown as one of the generic measures for assessed quality of voice. In this context service stability is understood to be the generic measure of incidence and persistence voice quality problems, as determined, for example, by the speed and reliability with which reported problems are corrected by the service provider. Table 9-2 expands on Table 9-1, by displaying the type of quality measures that are most appropriate for each of the objectives just described.

Alternatives for Quantification

Implicit in Tables 9-1 and 9-2 is the idea that a *generic measure,* or more simply, a *measure* is a description of some service attribute that can be expressed as a number or quantity. The measures here thus name *what* is being described, and *measurement* is the act of assigning some value to a measure. In this lexicon any expression or algorithm that details the calculations via which a value is to be assigned to a measure is referred to as a *quantifier* for the measure. Thus, for example, the

TABLE 9-1

Generic Measures of Service Attributes Affecting User Perception of Voice Quality

Type of Quality	Generic Measures
Intrinsic	Objective measures of end-to-end transmission characteristics, such as received speech power in dBrnC or a PESQ score
Perceived	Connection quality (MOS) Connection usability (P[UDI]) Expected incidence and severity of perceived impairments, such as speech distortion and echo

Assessed	Capacity requirements for acceptable voice quality
	Costs of acquisition of capacity
	  Initial
	  Continuing
	Service stability

TABLE 9-2

Utility of Types of Measures for Different Objectives

Objective	Type of Measure: Intrinsic	Perceived	Assessed
Network design	•	•	•
Market assessment		•	•
Service characterization		•	
Service validation	•		
Quality monitoring	•		

generic measure connection quality displayed in Table 9-2 is understood to be something to which a value can be assigned to gauge the likely user perception of what is heard. A class of quantifiers for connection quality are mean opinion scores, representing estimates of the average value that would be assigned were it possible to elicit and score opinions of quality of a service from the whole population of users. A specific quantifier in this class would be one of the following:

- A value determined from one of the many varieties of subjective tests that have been designed to sample such opinions
- A model that establishes the procedures for estimating a mean opinion score from other data elements

The reason for distinguishing measures and quantifiers in this way is that the quantifier concept fosters recognition and consideration of the fact that there may be many alternatives for assigning values of measures purported to represent the same thing. One benefit of such a realization is that it leads to the kind of distinctions described in Chap. 5, which avoid the pitfalls of assuming the existence of *a* mean opinion score. The more important benefit, however, is that different quantifiers for the same measure may reflect different effects and generate entirely different data-collection requirements. Consider, for example, availability of a service. Without the distinction between measures and quantifiers drawn here, the availability *A* is very likely to be defined by equating it to its most commonly used quantifier, setting:

$$A = \frac{\text{MTBF}}{\text{MTBF} + \text{MTTR}} \qquad \textbf{[9-1]}$$

where MTBF stands for mean time between failures and MTTR stands for mean time to restore. However, as a generic measure, availability is the probability that a particular service can be accessed and used at any particular point in time. Equation [9-1] is one quantifier of that measure, representing something that may or may not be a good estimate of the desired probability, depending on number of hours of operation observed. Equivalent quantifiers for availability are

$$A = \frac{T_o}{T_o + \sum_{i=1}^{N} O_i} \qquad [9\text{-}2]$$

and

$$A = \frac{T_{\text{tot}} - \sum_{i=1}^{N} O_i}{T_{\text{tot}}} \qquad [9\text{-}3]$$

where T_o = total time that service was observed to be operational
T_{tot} = total time service was observed
N = total number of failures experienced
O_i = duration of outage due to ith failure

Recognition of the equivalent quantifiers defined by Eqs. [9-2] and [9-3] thus suggest two alternatives for what data are to be collected, while resolving what are frequently uncertainties as to how to calculate MTBF and MTTR. Moreover, Eq. [9-3] for which the data collection is much more straightforward, makes it easier to adjust the estimate to exclude the effects of scheduled maintenance activity. If O_s is the total outage time due to scheduled maintenance, then the availability excluding scheduled maintenance can be obtained by replacing the denominator in Eq. [9-3] by $T_{\text{tot}} - O_s$.

Possible variations in effects reflected in quantifiers for voice quality are illustrated in Table 9-3, which shows the effects reflected in quantifiers produced from the various different tests and models that have been described so far. A cursory examination of that table shows, for example, that for a packet-switched transport:

- Voice quality evaluation aimed at understanding the effects of dropped frames can be based on any of the psychoacoustic tests or the VQES.

TABLE 9-3

Effects Reflected in Quantifiers Produced from Tests and Models

		Effect					
					Speech Distortion		
Technique	Quantifier	Speech Power	Noise	Echo	Encoding/ Decoding	Dropped Frames	Disruption of Normal Rhythms
PAMS, PSQM, and PESQ	MOS		•		•	•	
Electroacoustic analysis	Kurtosis of 2nd Δs		o*		•		
E-Model	MOS/ R factor	•	•	•	o†	?‡	•
VQES (or SAT)	MOS	•	•	•	•	•	
	P[UDI]	•	•	•	•	•	•

*Only when noise level is extraordinarily high.

†Reflects waveform distortion due to digitization; does not reflect transmission effects such as results of codec overdrive.

‡Planned for inclusion, but not implemented or promulgated as of July 3, 2002.

Note: • = fully reflected, o = partially reflected, ? = status unknown.

- Service analysis focusing on the likely effects of excessive round-trip delay must utilize the E-Model, the VQES, or some other multiple-effects model that includes round-trip delay as a factor.
- Service characterization should be based on the VQES or some other multiple-effects model that encompasses all the effects shown.

Note also that Table 9-3 emphasizes the pitfalls in misconception of MOS as a universal quantifier. The psychoacoustic techniques, E-Model, and VQES all produce MOS values, but with different effects reflected in the estimate. This means that even were those three MOS values on the same scale, they would represent entirely different ways of estimating what the user population would report for the service analyzed.

Table 9-3 exhibits a dichotomy that I believe is sine qua non for analysis of quality of packet-switched voice services—at least two independently quantified measures of perceived voice quality. This dichotomy is seen in the comparison of the VQES and E-Model. Where the E-Model produces a single measure of voice quality, jointly reflecting all the effects, the VQES produces two quantifiers, one designed to reflect only what is heard and another designed to reflect all effects on usability. An alternative pair of quantifiers comprising one that reflected all of the effects covered by the VQES MOS and another that reflected the conversational disruption would also be viable. However, the interpretation of a single measure reflecting both kinds of effects as is done in the E-Model must be handled with great care to ensure that the quantifier achieves an equitable weighting of their relative impacts.

Data Acquisition

The principal reason for identifying all possible quantifiers before attempting to collect the data for an analysis is that each quantifier defined specifies data elements that must be used in the calculation of values of measures. Such a specification defines in turn both what data must be collected and where the data must be acquired. For example, suppose a particular customer wants to see a comparison of voice quality based on the MOS derived from the PESQ. That stipulation automatically mandates collection of received copies of known speech waveforms transmitted across the service being tested. The data elements required are, then, recordings of the waveforms, which must have been acquired through active end-to-end testing.

Sources More generally, the principal data required for evaluating voice quality are images of voiceband waveforms as received at a termination point of the connection, or measures that are automatically produced through processing of such waveforms. The processing involved may be

- *Controlled.* Predicated on an assumption of the condition on the connection at the time the received waveform was captured
- *Comparative.* Requiring the ability to compare what was received with what was transmitted

Objective measures of voice connections commonly produced from controlled processing, include, for example, speech power, noise, and some electroacoustic measures of speech distortion. Those requiring comparative processing include signal attenuation, echo, and most psychoacoustic measures of speech distortion. These are the measures that are routinely captured and analyzed with telephone line test devices.

However, use of packet switching creates sources of impairments whose measurement may not be possible using the data-acquisition techniques employed for circuit-switched voice services. There are both controlled and comparative processing techniques for estimating dropped frame rates. However, quantification of intrinsic measures of quality of packet-switched services, such as packet latency, packet jitter, and round-trip delay represent problems of timing, rather than waveform capture. For example, measurement of packet latency requires data regarding when a packet left its origin and when it arrived at its destination, *synchronized to the same timing source.*

Fortunately, the importance of questions of timing and sequencing were recognized early on in the development of packet-switching technology, and most packet-switched traffic handling software provides, or can be readily configured to provide, capabilities to capture such timing data. For example, nearly all IP traffic handlers are complemented with routines for capturing and time stamping information in the packet envelopes. A typical routine of this kind can be implemented, for example, to capture the origin and packet sequence numbers from the real-time protocol (RTP) header and record the time of receipt. This capture creates a data set that can be analyzed to determine missing packets and packet jitter. Similarly, nearly all routers have implemented easily configured "ping" programs that can be exercised to sample the dropped packet rates and round-trip delay between packet-switching devices.

Such capabilities can be readily exploited to gather the kind of data needed for timing measurements. Acquisition of exactly what is needed may require a little invention and creative computing, but the potential to obtain almost anything necessary to complement the conventional telephony measurements is certainly there.

Means of Data Acquisition In the formal process described here, then, the selection of possible quantifiers will have determined what data might be necessary, and cognizance of the possible sources for the various data elements will have determined from where the data might be obtained. The final step in setting up a voice quality evaluation or service analysis will be determination of the means to be used to acquire the data. This step often results in a major benefit of the formal process, in that it frequently leads to the realization that there is an inexpensive, readily implemented alternative to the cumbersome, time- and personnel-consuming data-collection plan originally contemplated.

To assist in the selection of data acquisition means, the following provides a description of all the possible variants of data acquisition for a packet-switched voice service. The principal distinction in that profile is between active and passive data acquisition. We detail the principal alternatives within these classes and their characteristics in the following subsections.

Active Data Acquisition This type of data acquisition is termed "active" because it is based on transmission and reception of test signals that are carried across the service being analyzed. In the context of packet-switched services, there are actually two varieties of such active testing:

1. *End-to-end.* The first variant is end-to-end testing, in which the transmission originates at the same point that it originates for users and terminates to what would be the equivalent of a user's handset. The conventional options for such end-to-end testing for voice quality are subjective tests or objective measurements based on electronic transmission and/or capture of waveforms. There is in addition at least one device (see App. D) that will support hybrid testing in which the one side of the connection is a human and the other is an electronic device. With the development and deployment of telephone handsets designed to interface directly with a packet-switched network, the possibilities for end-to-end data acquisition will also include capture and processing of data in packet envelopes, such as that described earlier for RTP headers.

2. *Edge-to-edge.* A second variety of active testing that becomes possible in a packet-switched environment is what can be characterized as *edge-to-edge* testing, which can be used to acquire data between packet-switched network access and egress devices. Such testing is based on transmission of packets with known characteristics and capture and processing of the protocol data in their envelopes. An example is a ping test. The ping test is active, because it involves generation of signals to a destination requesting an acknowledgment. However, except for those cases where the device pinged is an IP telephone set, the ping test cannot be implemented end to end, because the ping request cannot be handled outside of the packet-switched transport network.

Passive Data Acquisition: Circuit-Switched Transport Passive data acquisition is also variously referred to as nonintrusive or in-service testing. The distinctive characteristic of passive data acquisition is that it is based on capture and processing of data on operational traffic being carried over the service being tested. For circuit-switched services, some data elements may be derived from capture and processing of messages exchanged in the out-of-band signaling systems that are used to route and set up calls. The capture of waveforms that can be processed to produce common objective measures is accomplished by use of low-impedance taps on digital transmission links that enable test devices to capture digitized signals as they are being transmitted without altering the signals or otherwise interfering with their transmission.

The possibilities for passive data collection for circuit-switched telephony are suggested in ITU-T Recommendation P.561, In-Service, Non-Intrusive Measurement Device—Voice Service Measurements (Ref. 8), which lists standards for such devices. That document specifies that the device must sit on a four-wire point in the end-to-end circuit and be capable of producing measures of

- *Speech and noise.* Speech level, psophometric weighted noise, and a speech activity factor for each direction of the transmission
- *Echo.* Speech echo path delay, and at least one echo loss measurement. The options for echo loss measurements are:
  - *Echo loss.* A quantifier of echo path loss that ascribes different weights to the loss in different portions of the voice frequency spectrum
  - *Echo path loss.* A quantifier obtained by determining the signal power loss without spectral weighting
  - *Speech echo path loss.* The echo path loss quantified on the basis of reflected speech signals

Other data-acquisition capabilities that might be implemented in these devices are displayed in Table 9-4, which lists the optional features mentioned in ITU-T P.561 (Ref. 8).

TABLE 9-4

Optional Features for In-Service Nonintrusive Monitoring Devices (INMD) Listed in ITU-T P.561

1. Originating and terminating address digits
2. Facility or circuit identification
3. Time and duration of connection
4. Signal classification (voice/data/other)
5. Customer identification (dedicated circuits only)
6. DS1 performance measurements
7. 3-kHz flat noise level
8. Connection disposition measurements
9. Data analysis and reports
10. Saturation clipping
11. Measurement interval
12. Double talk
13. Front-end clipping
14. One-way transmission
15. Crosstalk
16. Stability loss
17. Distortion

Because passive data acquisition obviates the need to place test subjects or electronic test devices at the sites served by the service to be analyzed, it is highly desirable when the data so acquired will be adequate for the objectives of the analysis. The problem is, however, that in many cases the data that can be collected are not adequate for accurately quantifying measures of perceived voice quality. The reason is illustrated in Fig. 9-1, which shows where INMD devices must sit relative to end-to-end telephone connections. As seen there, and listed in Table 9-5, there are a number of discrepancies between what can be measured at the INMD tap points and what users will actually hear. This means that passive data acquisition must inevitably rely on inferential processing of some kind in order to obtain values of measures that are commensurate with those obtained from active end-to-end testing.

Examples of the inferential processing challenges are illustrated by the following techniques that have been developed to use passively acquired data for voice quality evaluation.

- *Call clarity index (CCI).* The call clarity index is a multiple-effects model developed by British Telecom that was explicitly designed to support in-service testing of quality of telephone services. As described in Annex A to ITU-T Recommendation P.562, Analysis and interpretation of INMD Voice-Service Measurements (Ref. 9), this model converts the basic four INMD measurements—speech level, noise level, echo path loss, and echo path delay—into an estimate of a mean opinion score. The multiple-effects model

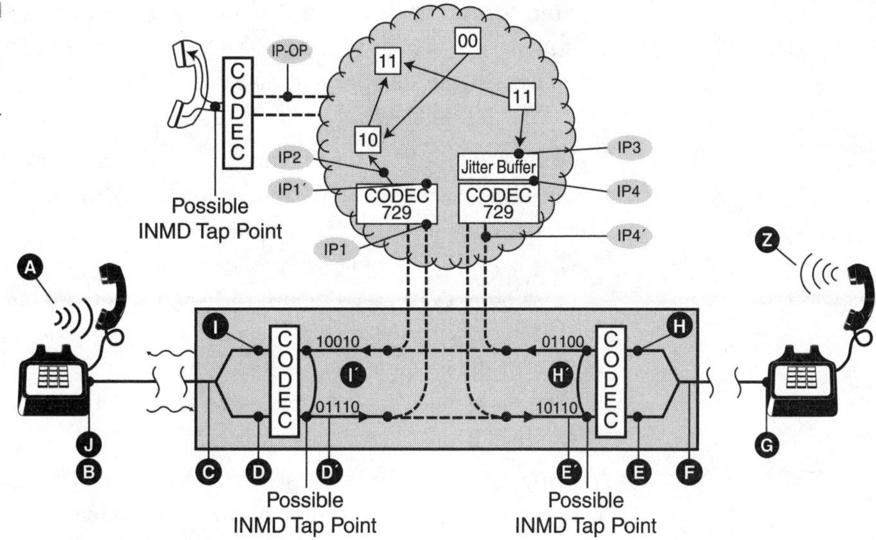

Figure 9-1 Access points for INMD testing.

TABLE 9-5 Discrepancies between INMD and End-to-End Measurements

Characteristic Measured	Included	Missing
Speech power	Origin side loop loss (B to C)	Transmission plan pad (E to F) Destination side loop loss (F to G)
Noise	Origin side line noise (B to C)	Destination side line noise (F to G)
Speech distortion	All effects	Nothing
Echo path loss	Transhybrid loss (E to H)	Transmission plan pad (I to C) Origin side loop loss (C to J)
Echo path delay (origin mouth to ear)	Near to far (E′ to H′) Far to near (H′ to E′)	Delays on origin loop—(B to C) and (C to J)
Round-trip delay	Echo path delay (C to J)	Delays on near loop—(B to C) and Delays on far loop—(F to G) and (G to F)

created for doing this is based on *assumed* values for no fewer than 12 characteristics of mouth-to-ear transmission, including 4 that are represented as spectral densities. The reason, as described in ITU-T P.562, is that

> To be able to predict the call clarity of a telephone call as perceived by the customers at either end the model requires the following information that is not available from INMD measurements:
>
> **1.** The overall sensitivity-frequency response characteristic of each transmission path (talker's mouth to INMD and INMD to listener's ear).
>
> **2.** The sensitivity-frequency response characteristic of each sidetone path (each talker's mouth to his own ear).
>
> **3.** The room noise spectra and levels at each end of the connection.
>
> (Reference 9, p. 15, paragraph 6.3.1)

- *Nonintrusive network assessment (NINA).* This is the trade name for a proprietary routine for estimating voice quality for passively acquired data developed by Swiss Qual, Inc. The estimate of a mean opinion score is produced by inferential processing of captured images of the received signal, without reference to what was transmitted. The processing includes, in addition to calculation of INMD speech level, noise, and echo measurements, complex algorithms for
  - Distinguishing periods when speech is present in the signal and separating them from periods when only noise or echo are present
  - Processing the speech periods to infer the characteristics of the waveform of the speech signal injected at the distant end
  - Processing the captured signal against the inferred speech waveform characteristics to detect, and correct for, instances of speech clipping and/or effects of dropped frames
  - Using the *inferentially reconstructed* injected speech waveform as the basis for comparison with the received waveform to produce estimates of the received voice quality

The inferential processing involved in some of these steps is probably even more complicated than it sounds, since the term *neural networks* shows up in their descriptions of NINA.

Passive Data Acquisition: Packet-Switched Transport In a packet-switched network, similar capabilities are afforded by extending the routines that support active edge-to-edge testing. This is accomplished, for example, by programming routes to look for and collecting the contents of packets carrying voice exchanges between terminals representing origins and destinations of interest. The RTP headers can then be processed to obtain data on dropped packets, packet latencies, and jitter. Moreover, given knowledge of the codec, codec configuration, and packetization scheme, the contents of the datagrams captured may be decoded to obtain PCM images of the transmitted voice signals. That can then be processed in the same way that signals captured with INMD devices are.

As suggested by Table 9-6, however, there is a major impediment to use of passive data acquisition for analysis of quality of packet-switched voice services. The problem is this. The voice signals heard by users are downstream from the codec used for the packet-switched transport, while the packet-switched data that can be captured passively are captured upstream of that codec. This means, for example, that the data on delay and dropped packets in the RTP header are not complete. There will be additional delays determined by the size of the jitter buffer and additional dropped packets that will result from delays that exceed the jitter buffer allowance. Moreover, the effects of the dropped packets on voice quality will depend on how the codec handles missing packets and is therefore manifested only after the codec has regenerated the transmitted voice signal.

As a consequence, the additional data that can be captured passively in a packet-switched network will be useful for purposes of evaluating voice quality or service analysis only if it is supported by inferential processing to estimate the downstream effects from the upstream data. The kind of processing necessary is suggested by the example of VQmon, a system developed to exploit passively captured data.

TABLE 9-6

Discrepancies between Passively Acquired Packet-Switched Data and End-to-End Measurements

Characteristic Measured	Included	Missing
Round-trip delay	Packet latencies in both directions	On both sides: Jitter buffer delays Codec encoding and packetization times Codec decoding times
Dropped packet rates	Packets dropped transnetwork	Packets dropped due to delay greater than those accommodated by the jitter buffer
Speech distortion, from reconstructed speech waveforms	Effects of Codec encoding and decoding Origin side noise	Effects of dropped packets, as determined by Codec configuration Missing packet handling

VQmon is the trade name for a proprietary extension of the E-Model developed by Telchemy, Inc., for nonintrusive monitoring of quality of VoIP services. It circumvents the data-acquisition problem by use of models that are applied to data captured upstream of the codec decoding to produce estimates of postcodec performance characteristics.

To accomplish this, the VQmon employs, among others, sophisticated algorithms to process data on packet timing and packets dropped across the packet-switched transport that can be readily captured from IP messages. The principal algorithms comprise

- *A jitter buffer emulator.* To determine from data on packet arrival times the additional packets lost because they arrive too late to be used
- *A family of codec models.* To determine how dropped packets affect the user perception of quality as a function of the type of codec and its configuration

These elements are, moreover, based on models that explicitly recognize the high volatility of IP transport characteristics. For example, in consonance with observations of IP performance, effects of dropped packet rates are derived from a model that assumes packet losses occur in bursts rather than at a constant rate. The effects of dropped packets on likely user perception of voice quality on a given call are then estimated in light of the *recency effect,* a documented phenomenon in which the same impairment occurring near the end of a call has a greater deleterious effect on user perception of quality than when it occurs earlier in the call.

The principal output of these processes is a codec-dependent *Ie* factor that can be used in the E-Model to reflect the effects of the IP transport on user perception of voice quality for each IP telephone call sampled. The effects of just the packet-switched transport on voice quality can be estimated by assuming the defaults for the other contributors to the E-Model transmission rating. Alternatively this *Ie* factor can be used in the E-Model in conjunction with separately measured connection parameters to produce estimates of voice quality end to end.

There will undoubtedly be other attempts to exploit passively collected data from packet-switched protocol messages. The important thing to recognize in this example, however, is that any system of this kind has to be based on something like the processing implemented in VQmon, *even when the passive data capture is implemented on the users' handsets designed to interface directly with the packet-switched network.*

Examples

To illustrate how the recommended analysis procedure works in practice, we conclude this chapter with the following descriptions of the application of the process to answer the final question: *What is the most cost-effective combination of data acquisition, handling, processing, and analysis to produce the required information?* The problems used as examples are ones, which, if they are not yet familiar to you, can be expected to become so as the push for development of packet-switched voice services begins to produce deployed systems.

Problem 1: Marketing's Nightmare

Your company is just about to release an announcement of a new packet-switched voice service. However, the marketing department has heard, through the grapevine, that the competition is ready to meet that announcement with an advertising campaign claiming that its service, which has been around longer, offers demonstrably superior quality. Marketing will not release the announcement until and unless you can give them ammunition to defend against the attack.

Analysis of Requirements The audience for whatever you produce will be the general public and/or technically naive national media representatives, who might be approached to "kill" the competition's ads, given credible evidence that the claims of superiority are exaggerated. The objective is to produce a credible comparison of your service and your competition's service expressed in terms that readily communicate to that audience.

This means that what you must produce is a service characterization. That characterization must be comprehensive, covering all aspects that users might experience with either service. The measures of voice quality on which it is based must, therefore, reflect both connection quality and connection usability. This immediately rules out any use of standard measures like the PESQ that do not reflect all possible effects on voice quality.

For something as all-encompassing as this, the alternatives are to do a lot of objective testing and use some multiple-effects model to produce estimates of subjective measures, or to execute a large subjective user test. If you pursue the multiple-effects option, however, you are not going to be able to use the model preferred for internal uses within your company. The intended audience will neither have the interest nor the patience to understand the features and virtues of that technique. Consequently, the only alternative to assure ready credibility and scientific defensibility of your voice quality evaluation will be to base it on the E-Model, or whatever other multiple-effects model that is currently endorsed by some authoritative body.

Whatever data must be collected to support the selected application, however, it is clear that the only data that will be appropriate for your purposes will be acquired via active testing of end-to-end connections. Anything less will expose your comparisons of quality to attacks on the efficacy and accuracy of the models used to transform passively acquired data.

Choice of Method The basic tradeoff to consider in this case is, then, between

- The greater cost of designing and conducting a well-designed subjective user test
- The threat that public arguments that might successfully impeach, or at least cast doubt on, a favorable comparison based on interpretation of objective measurements, even though the multiple-effects model used has been endorsed as a standard

Even given the predictable reaction of your company's management to higher costs, it is clear that you should advocate the conduct of a subjective user test. The compelling arguments supporting this conclusion are these:

- Because of the intuitive appeal and very high inherent credibility of comparisons of voice quality based on the assessments of unbiased persons using the actual services, a favorable comparison of voice quality based on a subjective test will finesse any claims of superiority based on any other kinds of testing.
- The claims of parity or superiority of your service supported by such a test will not be vulnerable to imputed weaknesses or limitations in the processing of the test data. In fact, when a subjective test is designed to elementary standards for sample sizes, statistical stability, and safeguards against biases, any attempt to argue against the results impugns the attacker rather than the test.

- The results of a large-scale subjective test will continue to have credibility unless there is some substantial change in the design of the services compared.

Thus, the subjective test may be more expensive in the short run, but there is much less risk that the results will not be usable for the intended purpose of gainsaying the competition's representations. Moreover, its results will be much more durable, much easier to communicate, and much easier to defend. This suggests that the initial expense will be offset by later savings in time and effort required to respond to proposals, formulate and defend advertising, etc.

Test and Analysis Procedure See, for example, Chap. 5.

Problem 2: The Designers' Dilemma

Your engineering department has come to the realization that at least some of the calls placed via a VoIP service designed to ride on the customer's intranet facilities must be completed to off-net destinations. There is so much uncertainty as to what might happen to the quality of the off-net calls that the company was tempted to tell the customer that off-net service would not be supported. Until, that is, they also realized that the VoIP service at many sites was slated to be a complete replacement for conventional services, using IP telephones instead of conventional handsets and a PBX.

To respond to this problem engineering has come up with the alternative of giving off-net traffic preferential handling in queues to reduce the exposure to unacceptable degradation of voice quality from what their users currently experience. The question to you, then, is whether that adjustment will be enough to prevent disaster.

Analysis of Requirements The objective in this case is to come up with a budget for the impact of packet switching on the quality of the off-net calls, so that engineering can determine whether the adjustment can satisfy that constraint. The audience comprises your engineers, who are conversant with telecommunications technology and are therefore competent to assess your findings without a lot of background.

As before, the measures used must reflect effects on both the quality of connections and their usability. The quantifiers in this case can, however, be almost anything that will establish a credible relationship between user perception of quality and the factors that the engineers are trying to control, namely packet latency, packet jitter, and dropped packet rates.

Because the decisions in this case must be made well before the customer's service is turned over for use, it is clear that a subjective test is not an option. This means that a robust multiple-effects model is the quantifier of choice. The fact that the model will be applied internally to produce results for a technically conversant audience also means that there is no restriction on what particular model should be used.

Choice of Method In accordance with the formal process recommended here, the choice of multiple-effects model to be used should be governed by examination of

- Alternative multiple-effects models to determine what data elements are required
- Readily available data, such as routinely collected measurements of circuit-switched connections to determine how easy it will be to produce the required data

Requirements for inputs describing the effects of packet-switched transport can, however, be ignored in this process, because the creation of the performance budgets will require hypothetical manipulation of those measures, rather than actual data. Given the various capabilities described in this book, this process would lead naturally to a selection of a multiple-effects model like the VQES, whose inputs comprise a relatively small set of commonly used measures of connections. For this application, this model would be complemented with something like the jitter buffer emulator described for VQmon to enable transforms of hypothetical values for packet jitter into the inputs required in the VQES.

Test and Analysis Procedures Given the choice of method here, the procedure would be straightforward. The readily available data on characteristics of the circuit-switched connections would be accumulated and processed to develop a profile of the quality of voice expected from the customer's current off-net services, as gauged by VQES MOS and P[UDI] outputs. The heuristics for acceptability of differences in these measures would then be applied to a range of hypothetical values of packet jitter, latency, and dropped frame rates across the transport to identify sets of values that would result in acceptable voice quality in the off-net calls. From that information your engineers would be able to determine whether the potential controls achievable with selective QoS would be sufficient to assure acceptability of the off-net service.

Problem 3: Your Customer Wants a Service-Level Agreement for Voice Quality

If anything will demonstrate the prudence of the analysis procedures recommended here, this problem will. The reason is that service-level agreements (SLAs) today are frequently negotiated on the bottom line, without any consideration of the questions that should be answered before undertaking any test and evaluation effort. As a consequence, the agreements often wind up being worse than worthless for the objectives of the customers who insist on negotiating them. This issue in measurement and evaluation of packet-switched voice services is so important that the exact step-by-step procedure is described here.

1. What Is the Objective? In simplest terms, the only reason for a customer's trying to negotiate an SLA is to foster actions on the part of the service provider that will reduce the frequency of, or ease the pain of, the unavoidable telecommunications service problems. The possible objectives in this regard, then, include the following.

Compliance with Specified Practices and Procedures. There is no way that a service provider can, or ought to, be held accountable for the consequences of uncontrollable events, or from conditions or events resulting from causes that both of you failed to anticipate. In lieu of such accountability, however, it is reasonable for customers to expect their service provider to adhere to design standards and operational procedures aimed at reducing exposures to such events. For example, where the service provider's equipment is powered by electricity supplied by the customer, the service provider cannot be held responsible for outages due to extended loss of power. However, the service provider may reasonably be expected to formulate procedures for reacting to such a condition to mitigate the effects of the power loss on telecommunications services and be held accountable for failure to implement those procedures in a timely fashion when needed. This suggests that one possible objective of the SLA might be to establish mutual understanding and agreement as to what practices and procedures the service provider is to adopt and adhere to.

Delivery and Maintenance of What Was Promised One of the major objectives of an SLA may be reassurance that the service delivered is, and continues to be, the service that the provider originally offered when the purchase contract was negotiated. This is, on the face of it, a very reasonable objective. However, its achievement in an SLA depends on how clearly the objective and its requirements for verification are understood.

Maximum Reasonable Effort Finally, for many customers, the thing that makes an SLA appealing is that it appears to create leverage for assuring that those deficiencies that do emerge will be corrected as quickly as possible. The rationale for this using an SLA to achieve this end runs something like this:

- There will always be transient episodes of unacceptably poor quality. Equipment will fail, software will malfunction, lines will get cut, etc. These kinds of problems are inevitable and unavoidable.
- When unacceptably poor quality is experienced, however, the amount of time that the condition will last depends on how the service provider reacts. When the problems are handled with dispatch and effective application of limited resources, they may be corrected before there is a material impact on the activities supported by the telecommunications service.
- An SLA will at least constitute a contractual agreement under which a failure on the part of the service provider will incur monetary penalties. That should provide some kind of incentive for better service.

Given such diverse possibilities, it should be clear that the first step in formulating an SLA must be to ascertain the customer's objective in negotiating an SLA for voice quality.

2. What Must Be Reflected in the Measures? Once the objective is identified, the requirements for what must be reflected in the measures defined for use in the SLA will be almost self-evident. For example, the procedures and practices that the customer may want to encourage will in this case be procedures for detecting deterioration of voice quality and the specific actions to be taken to react to such indications. The measures to support this do not have to be anything more than indicators of deterioration of voice quality, as might be derived from monitoring changes in service performance or the incidence of user quality complaints. The measures to be used to support verification of delivery and maintenance of the level of voice quality that was promised by the service provider will necessarily be those used in the service proposal and/or representations to the customer. And, an intent to foster maximum reasonable effort in maintaining voice quality on the part of the service provider will naturally require the use of a set of measures that reflects all possible deleterious effects on user perception of voice quality.

However, another aspect that must come into play in the selection of measures in this case is that the principal audience for the results of measurements is a customer who may have firmly held preconceptions as to how the SLA should be formulated. The request to negotiate the SLA may therefore include a specification as to what measure of voice quality is to be used, together with criteria for invoking penalties based on that measure. It will not be unusual, for example, to find that the customer has already decided that *the MOS* shall be *x.y,* and there shall be monetary penalties for any month in which that goal is not realized.

If the customer's perspectives seem to be well-informed or ignorant, but intransigently held, it is probably better in the long run to simply accept the customer's generic measure at this point and defer any education or refinement to the discussion of alternatives for quantification. The reason is that ambiguities and opportunities for confusion are generally much more readily resolved as the solution to a concrete problem of data collection and analysis than a clarification of an abstract concept.

3. What Are the Alternatives for Quantifying the Selected Measures? Depending on the measures selected, the options for quantification may run the gamut from direct application of simple calculations involving data acquired from end-to-end testing to application of very complex models that enable the use of passively acquired data. The only constraint in use of any of these alternatives will be what the customer will accept as reasonable and trustworthy. For this reason, the customer should participate in the examination of alternatives for quantification. An acceptable medium for such participation might, for example, be a conference in which the service provider would explain the alternatives for quantification, describing the strengths and weaknesses of each, showing relevant data comparing model estimates with results from subjective tests, and answering any questions that the customer might raise. The objective of this meeting would be to elicit customer acceptance of one or more of the alternatives to carry into the next step.

4. What Are the Options for Acquiring Data to Quantify the Selected Measures? The advantage of involving the customer in step 3 is that by the time the design of the SLA support system moves into the identification of data acquisition, the customer will have the same perspective of the utility of particular variants in data collection as the service provider. The customer will be able to see, for example, what the data from active end-to-end measurements will contribute to the reliability and ease of analysis as compared with applications of models to passively acquired data.

5. What Is the Most Cost-Effective Combination of Data Acquisition, Handling, Processing, and Analysis to Produce the Necessary Information? Actually, the real benefit from making the customer a partner in the formal process here is not so much the agreement as the method of voice quality evaluation. Rather, it is the concomitant understanding of the issues involved in assuring that the criteria for invoking penalties are meaningful and effective relative to the customer's objectives. By this point in the design of the SLA support system the questions of specific data elements and tradeoffs in cost of data acquisition are very concrete and easy to apprehend. It therefore becomes a relatively straightforward exercise to design the system to assure that the quantification methods that the customer has already accepted as being inherently credible will also be scientifically defensible, focusing on the following questions:

- What statistical tests will be applied to determine when a condition invoking a monetary penalty has occurred?
- What does that imply about the required minimum size and composition of samples?
- How much does it cost to acquire adequate sample sizes for alternative data elements?

What can happen is something like this: A customer starts out with the preconception that voice quality will be unacceptable to users unless the MOS as estimated by the ITU-endorsed PESQ is greater than 4.3. This position to that customer may be perfectly clear and unambiguous, suggesting that what is really to be accomplished in the SLA is a promise on the part of the service provider to deliver and maintain that level of quality. As any service provider will tell you, such a position is impossibly naive or ill-informed, because there are myriad problems in just determining what it means for the PESQ MOS to be 4.3 or greater. Does it mean that the 4.3 value must be achieved on every call? If not, does it mean that the average for all calls will be 4.3 or greater? If so, what sample size and statistic is to be used in testing for the average? Also, what is the expectation with respect to the way that the average is to be realized? There is, after all, a great difference in user perception of service in which a 4.3 average is realized from a normal distribution of MOS values and one in which 4.3 represents the average of calls of which 86 percent are "excellent" and 16 percent are unusable. Moreover, as was pointed out earlier, PESQ does not reflect all the effects that may degrade user perception of voice quality; it is possible to have a completely unacceptable service for which the PESQ-based MOS is even better than 4.3 for all calls.

To try to tell that customer any of this at the outset is to invite derision as being evasive about the quality of your packet-switched voice services, because unscrupulous providers are lining up to promise just that and move on, knowing full well that the customer will never be able to gainsay the claim after the service is installed. Rather, by following the analytical process recommended here, the transform will begin with the examination of possible objectives. In the process of eliciting a clear understanding of the customer's objective, it will be completely proper to ask such telling questions as: "Do you mean that you want us to respond quickly to voice quality problems only when and after PESQ measurements reveal a drop below 4.3?" Moreover, two of the three objectives that might eventually be agreed upon will finesse the issue of basing criteria for invoking penalties on PESQ values, because the measure required is something that can be rapidly interpreted to detect a *change* in voice quality, rather than an actual subjective measure.

However, even assuming that the customer wants the promise of a PESQ-based MOS of 4.3 or better, and the objective is to verify that what is promised is delivered and maintained, the other steps will lead the customer away from the original position. Requirements for direct quantification of the PESQ will dictate much more costly testing, due to license fees for the algorithms and necessity of large-scale active end-to-end testing to use the PESQ. The examination of less costly and more readily handled alternatives for acquiring data to quantify the MOS will eventually reveal that the PESQ is more costly and less effective than direct quantification and interpretation of the connection characteristics that determine the MOS value that will be produced by the PESQ.

It has, in fact, been the unstated objective of this part of this book to equip its readers with the insights, perspectives, and knowledge necessary to carry something like this off. If it seems like this is a reasonable, achievable scenario, then that goal has been realized.

PART 3

Other Aspects of Quality of Service

The discussion of challenges of measurement and evaluation of packet-switched voice services has so far focused almost exclusively on the problem of gauging likely user perception of connection quality and usability. While this is the central issue in assessment of effects of packet switching on quality of voice services, it is by no means the only aspect of QoS that may be changed with the transition from today's circuit-switched services.

Accordingly, this section of the book turns to an examination of those other aspects of QoS. Unlike the issue of voice quality, however, there are no questions as to what to measure or how to evaluate the results. Rather, the open issues in this case are how well-known, widely accepted metrics will be affected by the setup and handling of packet-switched calls. Such questions must, moreover, remain open until as yet undecided specifics of routing and monitoring packet-switched calls are resolved. For example, there will be no way to estimate the effects of IP call handling on routing speeds until there is a resolution as to whether the session initiation protocol or H.323 or both will be used or whether some other call origination protocol is going to be used.

It is, therefore, impossible at this time to present anything substantive on this topic beyond the warnings presented here as to what to look out for as packet-switched voice services become operational.

CHAPTER 10

Other Quality of Service Concerns

The Quality of Service (QoS) Model

The question of voice quality is a dominant, but by no means exclusive, concern that users express when they ask for reassurances that the quality of their voice services will be acceptable. A comprehensive description of what determines users' perception of quality of their telephonic voice services includes no fewer than six independent service characteristics besides the voice quality that has been the subject of Part 2. The principal characteristics are described in Ref. 1 as being

- *Accessibility,* the ability to initiate a call when desired
- *Routing speed,* the speed with which calls are set up
- *Connection reliability,* the reliability of the process of setting up a call
- *Routing reliability,* the reliability of the process of routing the connection to the requested destination
- *Connection continuity,* the ability to maintain a connection with acceptable quality until it is no longer needed
- *Disconnection reliability,* the reliability of the system responses to the instructions to cease, and therefore stop billing for, the connection

Internet zealots will immediately object to the names given these aspects of perceived QoS, because Internet voice services are connectionless. However, the descriptions of each that follow should clearly define exactly what is involved, even for "connectionless" connections.

Accessibility

Accessibility is the dimension of perceived QoS that reflects the reliability of the system providing the service as manifested to its users. The basic user concerns with accessibility are invariably expressed by questions such as

- Will I be able to place a call whenever I want to?
- If not, how long will it be before I can?

The generic measure of accessibility is, then, defined by

1. Agreeing that an *operational service interruption* is something that prevents use of the voice service during periods that users expect to

be able to access the system (e.g., by getting a dial tone or other confirmation that a call can be placed)

2. Defining the accessibility to be AC[*t*], the probability that a user attempting to originate a call will encounter an operational service interruption lasting *t* time units or longer, where the value AC[0] is understood to represent the probability that no service interruption will be encountered.

Routing Speed

Given access to the service, the next step by the user is to specify the call that is to be set up. Routing speed refers to the time it takes for the system to respond to that specification. User concerns with routing speed are expressed in questions like

- How long does it take before I know that the call has been extended to a station that might be answered by the party I have called?
- Is that time stable and predictable?

Readers familiar with Ref. 1 may immediately note that the first question here is expressed more generally than its counterpart in that book. The reason is that packet-switched voice systems may support placement of calls to persons, rather than stations, and be routed by cascading follow-me protocols in which alternatives are tried without the users' redialing.

The basic quantifier for routing speed as perceived by the user is the *postdial delay,* defined in most general terms as the time lapsed between the user's input of the last piece of information needed to define what connection is desired and the receipt of the first indication from the system of the disposition of the request (e.g., indication that the call has been set up, can't be set up because the station is already in use, could not be set up because of a routing problem).

In today's circuit-switched services, for example, the postdial delay is the time between the last digit dialed and reception of the first audible response from the network, which may be

- *Ring back.* Indicating that the call has been connected to a distant station
- *Station answer.* Indicating that the call was connected
- *Slow busy signal.* Indicating the called station was contacted but found to be busy

- *Fast busy signal.* Indicating that the call setup attempt was abandoned
- *Special information tone and/or recorded voice announcement.* Indicating that the requested call could not be set up and/or why

As will be seen, there is a real question as to what will be the counterpart of these responses in a packet-switched voice service.

Connection Reliability

Connection reliability is the attribute of a telephone service manifested to users by return of an expected response to a request to set up a call, indicating that the call setup attempt was not thwarted by some systemic malfunction. User concerns with such connection reliability are most frequently expressed by the colloquial question: Will my calls go through the first time I dial?

More formally, the generic measure of connection reliability is the probability that a correctly executed request for a call setup will be extended all the way to a destination station. As manifested to the users of circuit-switched services with common network responses, for example, the appropriate quantifier for connection reliability is the *normal completion rate* (NCR). This measure is defined to be the expected proportion of call attempts that result in normal completion indicators, comprising ringback with or without an answer, a slow busy signal, or an answer from a distant station.

Routing Reliability

As used in this model, routing reliability refers to the accuracy of call setup. The user concern is expressed as the question: When I request a connection and one is set up, can I be sure that it has been set up to the destination I wanted? Underlying this question is the user's recognition of the possibility of *misroutes,* and the easy quantifier is the expected proportion of properly executed call requests that will be completed to and answered by the wrong station.

In conventional voice services, such misroutes rarely cause any greater problem than the minor inconvenience of having to apologize for error. The calling party reaches an unintended party; one of them says, "Sorry, wrong number"; and all is usually forgiven and quickly forgotten.

As will be seen, however, packet-switched voice services offer possibilities whose consequences may be much greater than this.

Connection Continuity

Connection continuity refers to the ability to maintain a connection in good condition until the transaction initiated over it is satisfactorily completed. In the case of voice, the continuity of a connection thus refers to the ability to maintain an interactive exchange of information until at least one party is ready to end it. The user concern with connection continuity is colloquially expressed by the question: Once we start talking, will we be able to stay on the line until we are through?

The conditions that might result in discontinuities in voice connections include

- *Spontaneous disconnects.* The call that was set up is torn down without a request (on-hook action) by either party.
- *Unacceptable degradation.* The conditions on a connection that started out with acceptable quality become so bad that both parties agree to an early termination of information exchange.
- *Errors in transferring a connection.* A disconnect results rather than the intended action.

The obvious quantifier is the expected proportion of calls that result in such early disconnects, expressed as a function of the duration of the call.

Disconnection Reliability

The final aspect of quality of a voice service as perceived by its users is the reliability with which instructions to disconnect a call signaled by an on-hook condition produce the desired result. Problems with disconnection reliability affect voice users only when that failure to tear down a call when requested results in billing for an inordinately long call with an inordinately large charge attached. When this does happen, however, neither the customer nor the service provider is really going to be happy with the necessity to correct the billing. Disconnection reliability, expressed as the expected proportion of calls that are torn down within a few seconds of the transmission of an on-hook signal by either party may, therefore, represent a very important measure of quality of service.

Call Routing across Packet-Switched Networks

In simplest terms, telephone calls are set up across a circuit-switched network in four easy steps:

1. *Registration.* The user takes the handset off-hook, waits for a dial tone, and dials (or more properly nowadays, punches in) the number of the station to which the call is to be connected. As those numbers are dialed, they are transmitted to a local telephone switch in the form of particular tones that are detected and interpreted by the switch as digits. [The tones comprise two distinct frequencies for each number. They are referred to technically with typical redundancy as dual-tone multiple frequency (DTMF) tones.]

2. *Routing.* The digits captured are forwarded along on lines connecting possibly several intermediate switches until they are presented to a switch in the out-of-band signaling network. When the digits arrive there, they are translated into a request for finding a series of switch to switch links that can be interconnected to set up a connection to the local switch at which the station called is terminated. This process takes place very quickly, via an exchange of messages transmitted through the out-of-band signaling network. The messages are sent switch by switch to verify the availability of, and reserve, an interconnection between an incoming line and an outgoing line across each switch.

3. *Cut through.* When the request for connections is finally completed, allowing for the set up of a link-by-link connection from the local origin switch to the local switch serving the destination, the reserved connection is set up. At this time, the local switch is fed the digits identifying the line to the destination station, and the final interconnect is attempted.

4. *Notification.* The results of a final interconnect attempt are then signaled back to the origin station via either a station busy signal, indicating that the terminating line is already in use, or a ringback, indicating that the connection is set up as far as it can go and will be completed for conversation when the called party picks up the handset. This notification is sometimes preempted by an answer on the line before the first ringback signal is generated, in which case the answer serves in lieu of the notification.

No matter how it is implemented, routing via packet switching is going to add two new steps to this process:

1. *Registration translation.* The digits in a telephone number are uniquely associated with an electrical connection between a local switch and a particular telephone station. However, in a packet-switched protocol the analogous addresses are uniquely associated with a particular device, which may be located almost anywhere in the network at any time, or are aliases associated with a particular person that can be translated to be the address of one of a number of devices, depending on which is active at a particular time. This means that a packet-switched telephone service must be supported by a database from which a standard telephone number can be associated with addresses of devices served by the packet-switched network. In a hybrid voice service, for example, a call originated through a local switch must be routed to an edge device in the packet-switched transport network. Upon being announced at that device, the telephone number must be translated into the address of the device via which the call will exit the packet-switched network. If the exit is another edge device linked with a telephone switch, the number dialed must be communicated back to the out-of-band signaling network. There may, therefore, be as many as two translations, number to address and address to number, required to relay the data necessary to set up the call to the distant station.

2. *Termination negotiation.* No matter how the routing is accomplished for other steps of the call setup process, traversing the packet-switched transport requires the assistance of an intermediary device to negotiate the communications between the entry and exit devices. The options regarding how to effect such negotiation are many, varied, complicated, and currently the subject of internecine warfare among the advocates of different protocols. Those interested in the gory details of all this for the Internet will find more information on the topic than one might believe could exist in Ref. 10. For our purposes, however, it suffices for you to understand that the basic negotiation takes place as a dialog through an intermediary device that runs in machine talk something like the exchanges shown in the box in Fig. 10-1.

Implications for QoS

As suggested by the preceding, use of packet switching for voice transport will affect all six aspects of quality of service defined there. As described here the effects will be attributable to

Figure 10-1 A stylized representation of termination negotiation across a packet-switched transport.

A recently intercepted termination negotiation between Gatekeeper and Eager Caller ran something like this:

Eager:	Keeper, this is Eager. I need to get a call set up with Called.
Keeper:	Easy there, Eager. I've got other calls waiting ahead of you. I've got to get them started before I can take care of yours. And you'd better have the right addresses for me when I do. I'm a Gatekeeper, not a Gateway, you know. I don't work with these silly numbers. So give me what you want, and I'll work it for you.
Eager:	I want to set up a call with Called. I'm at the end of a 128-kbit/s connection, and I'm running a G.729A-VAD voice codec. My net card address is 198.735.978.7654.
Keeper:	Okay, okay, Eager. I got it. Hold on for a moment.
Keeper:	Calling Called. Calling Called. This is Keeper. Please come in. I have a call for you.
Called:	Hello, Keeper. I'm out here on domain address 234.679.998.5437. Who's calling?
Keeper:	I have a call request from Eager. Says he's real interested in getting to you. Wait one.
Keeper:	Eager, I'm in touch with Called right now. Hold on just a sec.
Keeper:	Anyway, Eager wants to come to you via G.729A-VAD. What say?
Called:	Oh, man! No can do. Tell Eager that it's G.729 with no VAD or *nada*.
Keeper:	Eager, Called sez you gotta come to him G.729 with no VAD or it's no dice. Can do?
Eager:	Oh, what the heck. Yea, yea. Tell him okay that way.
Keeper:	Called, Eager says to come on with your G.729. His address is 198.735.978.7654. Put that in your router and answer back directly. I'll tell Eager that you are on the way, so he'll be looking for you.
Keeper:	Eager, I got Called. He's coming to you as domain address 234.679.998.5437. Set up the ringback signal to the handset you're routing for, and go listen for him.
Eager:	Keeper, I gotcha. I'm on it.
Eager:	Hey there, Called! I got your message. Let's talk!

- The characteristics of hardware or software involved in the use of packet switching
- The additional routing functions necessary to make a packet-switched telephone service work like the circuit-switched one to which its users are accustomed

They will, in addition, depend on whether the kind of packet-switched voice service involved is hybrid transport or packet-switched telephony.

Whatever the source of the effects and service type, however, there is nothing among them that can be expected to improve the associated measures of perceived QoS for voice services. Rather, the real question is how great those effects will be, and that cannot be resolved until some of these systems are deployed for user service so that the expected effects can be measured.

Hybrid Transport

In a hybrid transport system, call origination or termination, or both, are effected via circuit-switched facilities. The packet-switched transport therefore serves as an intermediate relay service whose use requires a number of functional accommodations to handle transmission into, out of, and across the packet-switched network. Although one side of calls across a hybrid transport may terminate into a packet-switched telephone, in our description of possible effects here we will assume the worst case of circuit-switched call origination and termination.

Accessibility One of the features of today's circuit-switched long-distance networks is that they are *multiply connected*, offering many different switch-to-switch links that might be used to set up a particular connection across the network. As a result, the accessibility of the network is determined almost entirely by the availability of access and termination facilities. In the case of a residential or small business service this means that the accessibility is determined by the availability of the subscriber loop that connects a handset to the local switch during periods when users are present. For the case of customers with a large number of users served at a given site, the access is created by an arrangement of private lines routed around the local switch to terminate directly into the long-distance network. The accessibility of such a virtual private line service then depends on the diversity and redundancy incorporated into the service access design. When the accesses are all carried on one transmission facility into one switch, the service is said to be *single-threaded,* and access will fail any time the customer PBX, the terminating long-distance switch, or the transmission facility connecting them fails. For that reason, large customers will usually insist on access arrangements whereby the entries into the long-distance network are set up through more than one long-distance switch (diversity) and use different transmission facilities between them (redundancy).

A packet-switched network is designed to afford the same multiplicity of connections over which the individual packets can be transmitted from their origin at the edge of the network to their destination. As a consequence, there is very little difference between the inherent accessibility of the packet-switched and circuit-switched network.

In a hybrid network, then, accessibility will be degraded, to the extent that *the connections from the circuit-switched network into the packet-switched network and out of the packet-switched network into the circuit-*

switched network have less diversity or redundancy that the analogous multiple connections within the long-distance network. In practical terms, this means that if you are looking at a long-distance network in which every switch connects directly to at least three others, and hybrid design calls for only two different gateways into or out of the packet-switched network from each switch, beware! You may be buying into a noticeable difference between the current circuit-switched service and the hybrid.

Call Setup Characteristics Collectively, the routing speed, connection reliability, and routing reliability gauge the likely user perception of quality of a telephone service's handling of call setup. As detailed in what follows, the deleterious effects of the use of packet switching in a hybrid network on these characteristics may be substantial.

Routing Speed In the case of routing speed, the principal effect of packet switching will be an increase in the postdial delay created by

- Execution of all the translations needed to maintain compatibility of the circuit-switched and packet-switched addressing
- All the exchanges needed for termination negotiation

Specifically, where the expected delay is 0.2 to 0.6 s needed for routing a call across the long-distance network, the routing across the packet-switched network will add the time required for

1. *Translation of dialed numbers into device addresses and back.* If this is accomplished in the hybrid network via database lookup the same way that it is in today's circuit-switched network, the additional time can be expected to be about 0.5 s per translation. The translation from, say, a 10-digit North American numbering plan into an IP address on entering a VoIP service network and from the IP address back to the destination number might, therefore, be expected to add about 1 s to the postdial delay.
2. *Negotiation of termination.* Figure 10-1 presents a stylized representation of the fewest exchanges required to set up a call across a packet-switched network in accordance with the protocols most often proposed for this application. Even that minimal requirement represents at least 16 packets that must be transmitted, received, decoded and have a responding packet generated. With a nominal packet latency of, say, 50 and 5 ms, respectively, for reading and writing the response, this would add about another second to the postdial delay.

Thus, even very crude estimates suggest that the additional handling for routing a call across a packet-switched network without setting up "tunnels" or virtual channels may be substantial at best, and unacceptable at worst. Again, this does not say that it has to happen, only that the service quality analysis for hybrid services should recognize the potential and make provisions for verifying effects on postdial delay.

Connection Reliability It is clear that the additional handling involved in routing calls across the packet-switched transport will degrade connection reliability. Sporadic failures in the processing and message exchanges implementing registration translation and termination negotiation will inevitably increase the proportion of call attempts not completed. Although the size of the effect cannot be readily estimated without an opportunity to test an operational hybrid transport service, it is safe to assume that the incidence of such failures will be at least as great as that experienced in the out-of-band signaling system in the circuit-switched network. If so, expected call completion rates on the order of 99.5 percent in the circuit-switched network will drop to a marginally acceptable 99.0 percent in the hybrid network. Such an effect is rife with possibilities for dissatisfaction.

Routing Reliability The same observation holds for routing reliability—the additional processing involved in routing calls across the packet-switched network will probably at least double the incidence of misroutes due to errors or faults in the packet-switched call handling. Moreover, there will also be an additional source of misroutes in this context, in that the features and functions envisioned for telephony employing packet-switching technology will create opportunities for routing mistakes that simply do not exist in today's circuit-switched services. For example, "soft-switching" through a packet-switched network may support the use of aliases for multiple devices in the network associated with a particular person, comprising, say, telephones, a palm pilot or pager, a desktop computer, and a laptop computer. (Soft-switching is routing end-to-end connections by use of the packet-addressing protocol rather than specification of node-to-node connections.) Moreover, the alias may refer to a number of different telephones that a user has located at different sites, some of which are IP and some that are conventionally terminated. This possibility creates what is, in essence, another dimension of routing reliability. The caller's concern will be "Was the call routed to the correct device serving the party I called?" From the perspective of the called party, the concern will be, "Was the call routed to a device at which I did not want to receive a call of that type?"

These new concerns become particularly important for a general-access hybrid voice service via which calls out of the PSTN may be routed to multiple terminations via the packet-switched network. They may, in fact, come to be the dominant concerns that must be addressed with measurement and evaluation of routing reliability.

Connection Continuity As indicated earlier, there are three principal sources of failures of connection continuity: spontaneous disconnects, not requested by either party; unacceptable degradation of initially acceptable quality; and errors in transferring a connection. Of these, the incidence of spontaneous disconnects and transfer errors will have to be somewhat higher than that for a circuit-switched service due to the additional call handling involved in the packet-switched network, but not necessarily noticeably so.

However, the incidence of early disconnects due to degradation of quality is sure to be significantly, and probably noticeably, greater in a hybrid transport service. The reason is that voice quality problems are manifested differently in circuit- and packet-switched networks. In a circuit-switched network, a poor-quality connection is evident as soon as there is an exchange of information across it. For this reason, the early disconnects due to problems in the circuit-switched transport will not be affected by the intermediate packet-switched transport. However, calls connected across a packet-switched transport can start with acceptable voice quality and later experience a rapid deterioration of connection quality or usability due to transient congestion effects. This means that the expected incidence of early disconnects made because of unacceptable voice quality will be the sum of

- All such disconnects expected in the circuit-switched network
- All those expected due to transient deterioration of quality in the packet-switched transport

This suggests that a hybrid transport service is very likely to be perceived as having a noticeably inferior connection continuity than its purely circuit-switched counterpart.

Disconnection Reliability In a circuit-switched network the request for disconnection is signaled by the act of placing the receiver on-hook. This act opens (causes a drop in current flow across) the subscriber loop that was activated when the handset was lifted. The change in state is detected at the local switch. In response, it disconnects the incoming line, which sends a similar disconnect signal back to the previous switch in the chain of interconnections by which the call was set up. This continues until the indication that the call was disconnected is detected by the first switch in the out-of-band signaling network, at which point that system takes over, sending specific messages as to what connections to drop. At the far end of this process, where the call originated, the drop connection instructions eventually revert to electric signals that cause the loops to open.

Now, this process in a circuit-switched telephone network is very reliable. There are, nonetheless, sporadic instances in which the disconnection process fails, for example, because the on-hook action fails to open the subscriber loop or the instruction to take down the connection is not properly forwarded to the out-of-band signaling system. Such failures are partially manifested to users by such abnormal events as

- Picking up the telephone after talking to someone and discovering that the call is still connected to the distant end
- Being informed that you didn't hang up by that loud "wah-wah-wah" signal or the famous voice announcement, "If you'd like to make a call, please hang up and try again."
- Appearance of a charge for a call that lasted 17 h on the long-distance bill!

The analogous process of disconnection of "connectionless" connections in a packet-switched transport will be implemented via an interactive exchange of messages among various devices, something like that stylized in Fig. 10-2. Although the exchanges internally among the gatekeeper, the gateways, and the routers are not that much different from those necessitated in the out-of-band signaling used in the circuit-switched network, there is the added element at the gateways of

- Recording the associations between circuits and routers and other assigned handling devices for all connections traversing the interface between the circuit- and packet-switched transports
- Keeping track of origin, destination, and configuration of all active sessions supporting packet-switched transport of circuit-switched network traffic
- Effecting proper translations for messages that have to be relayed with the same content both internally, to packet-switching devices, and externally, to the out-of-band signaling system

E-Router:	Keeper, this is Eager's Router. I've just received an end-of-session message for the voice data out of 198.735.978.7654.
Keeper:	Okay. Let me see here. The time is...the net card is...Oh! What's the basis for the request to end the session?
E-Router:	It looks like a normal disconnect signal received and translated at the Gateway.
Keeper:	Well, if that's what they want...Wait one.
Keeper:	C-Gateway, this is Keeper. We have a valid disconnect request for 198.735.978.7654 out of E-Router. The destination is 234.679.998.5437 out of C-Router. Can you confirm routing?
Gateway:	Keeper, this is C-Gateway. I confirm. I say again, I confirm E-Router origin and C-Router destination.
Keeper:	Okay. Wait one.
Keeper:	E-Router, I confirm receipt of disconnection request from Eager. Send your session termination message.
Keeper:	C-Router, this is Keeper. Please generate a session termination message for the session active through you to 234.679.998.5437 and confirm.
E-Router:	I confirm transmission and receipt of session termination message for 198.735.978.7654, which is the access for Eager Caller.
C-router:	Keeper, I confirm transmission and receipt of session termination message.
Keeper:	E-Gateway, this is Keeper. All disconnect requirements executed. Delete association and free endpoints.
Keeper:	C-Gateway, this is Keeper. Delete our earlier-referenced association and free your router slot.
E-Gateway:	Keeper, it is done. Session has been logged out and a positive response to the termination request has been sent back to Out-of-Band on my side.
C-Gateway:	Keeper. Done. Order to disconnect has been sent out to Out-of-Band on my side.

Figure 10-2 A stylized representation of disconnection negotiation across a packet-switched transport.

As is the case with routing and connection failures, the double-duty processing involved here suggests that the incidence of disconnect failures for calls completed across a hybrid transport service might be as much as 2 to 4 times greater than that for its circuit-switched counterpart.

Packet-Switched Telephony

As just described, the major concerns with QoS of a hybrid transport service devolve to the question of how closely the hybrid system will resemble its circuit-switched counterpart. In order to make the combined use of two different systems look like one to the user, it is necessary to invoke additional call processing and handling that is not required in either. As a result, the vulnerabilities to problems affecting accessibility, call setup characteristics, etc., become additive, and the issue becomes whether the overall effect on measures of quality will be great enough to be noticeable to users.

In the case of packet-switched telephony, the interfaces between circuit- and packet-switched services described in the previous section are not required. The question of adequacy of quality of service therefore reduces to one of how measures of quality for a particular packet-switched telephone service compare with those for an equivalent circuit-switched service. There may be some differences in the quantifiers for these measures due to idiosyncrasies in the event monitoring under the packet-switching protocol used. However, if the quantifiers used for packet-switched services are carefully chosen to be a good estimate of the generic measure, the resultant values should be commensurate.

To close this examination of issues of measurement and evaluation of the other aspects of QoS, the following very briefly summarizes possible differences between circuit- and packet-switched telephony that might have to be taken into account in defining appropriate quantifiers of this kind. For purposes of illustration, the discussions presume that the services being compared are

- A virtual private network arrangement utilizing direct accesses and terminations to a circuit-switched long-distance network
- A packet-switched service utilizing IP telephones to carry VoIP over the company intranet serving the same sites.

Accessibility Because the circuit-switched long-distance and intranet services both provide a transport that is for all practical purposes able to support continuous, totally uninterrupted ability to effect edge-to-edge communications, there is a temptation to ignore accessibility. Such a presumption of equivalent accessibility must, however, be validated by an examination of the expected availability of the accesses to the edge. A single-threaded circuit-switched access will be configured with

- A PBX to handle internal call origination, route the call, and provide the routing data to the long-distance switch
- A channel bank or equivalent device to digitally encode and decode incoming and outgoing signals and inject them into a multiplexed high-speed digital transmission line into the long-distance switch
- The long-distance switch itself

Since the accessibility of a multiple-access arrangement will depend on the availability of single-threaded connections like these, the principal contributions to accessibility that must be taken into account are the expected frequencies of, and average time to recover from, failures of these components.

A single-threaded intranet access arrangement for the IP telephone will generally involve

- The IP telephone itself, which will generate the call setup signaling and digitally encode and decode the voice signals
- An Ethernet connection to a local server that will multiplex the traffic into a high-speed data transmission line into a router
- A router that will direct outgoing packets to their destination and feed incoming packets into the appropriate server
- A firewall device that will check what is going into and out of the router to verify that the origins and destinations are valid devices attached to the company's intranet

Given the obvious differences between devices and transmission lines, it will be necessary at least to compare the availability of the single-threaded accesses before concluding that there will be no significant difference in accessibility. Moreover, where the circuit-switched service employs redundancy and diversity to achieve better accessibility, it will be necessary to verify that the access configurations via the intranet have been, or can be, configured to achieve comparable levels.

Routing Speed Even though the Internet protocols will support capture of myriad details on the progress and outcome of call setup, the quantifiers of *perceived* routing speed must be predicated on what users hear. This means that accurate estimation of the analogue of postdial delay in the packet-switched environment will require information on the content and time of detection of audible signals that will replicate, or at least serve in lieu of, those generated today. Thus, even though it will be possible to routinely capture the time of transmission of the first call initiation request message and the time of receipt of the announcement of recognition of the call by the distant end, the difference in these times may not accurately reflect the user perception of routing speed. Rather, the delay calculated from these times must be adjusted by an estimate of the *ring latency*, which represents the time between call setup and the time the audible ring signal is presented to the person originating the call. This is no different from the adjustment that must be made when estimating the postdial delay from the call setup time in a circuit-switched environment. However, the ring latency in this case will have the unusual property that it may be negative under some protocols for VoIP, when the message to start the ring signal precedes the attempt to elicit contact with the device called.

Connection Reliability The idea of quantifying perceived connection reliability as a normal completion rate, representing the proportion of outcomes of call attempts that are seen by users as the expected outcome of a successful connection attempt is straightforward enough. However, until we know what the audible responses generated are going to be, what they mean, and what amplifying information might be displayed on the IP telephone screen, we will not be able to determine if IP telephone users are going to discriminate between normal and abnormal outcomes. For example, today a ring-no-answer condition is perceived to be a normal completion, indicating only that the party called was not available to answer the phone. If that ring-no-answer condition is augmented by a message on the IP view screen stating "Session request rejected at handler," there may be a slight modification of the user's interpretation of the ring-no-answer condition.

Routing Reliability Although the generic measure may be the same, the whole focus on routing reliability changes in a VoIP service. Since IP services are known to be vulnerable to any number of different kinds of attacks aimed at intercepting message exchanges, denial of service with spurious traffic, etc., the basic concern is, "When I request a connection and one is set up, can I be sure that it has been set up to the destination I wanted?" This basic concern is amplified, for example, with

"What is the chance that my voice signals will, without my knowledge, also go to a destination where they can be heard by someone I really would rather not hear them?"

"What is the chance that the person I am talking to is not the person I think I called?"

In practical terms, this means, on the one hand, that the quantifier for routing reliability in this context must recognize and reflect the possibility of deliberate routing errors. On the other hand, the additional concerns suggest that the evaluation of a particular value of routing reliability may be different—users exposed to more than inconvenience will naturally demand much higher routing reliability than today's users of circuit-switched services.

Connection Continuity The quantifier for comparisons with circuit-switched service must in this case reflect an accurate estimate of the expected incidence of early disconnects due to transient poor quality experienced after the call has been completed. The reasons were explained earlier in the discussion of connection continuity for hybrid transport services.

In addition, many varieties of IP telephones will implement directly many of the call-handling features, such as hold, call waiting, call forwarding, and call conferencing, that are currently implemented in PBXs. The error rates for these actions in the two systems should be estimated and compared as a measure of the relative usability of their different implementations.

Disconnection Reliability The appropriate quantifiers for *perceived* quality of service for the IP telephony are in this case exactly the same as those for circuit-switched telephony. The major possible difference is that VoIP protocols may support much easier capture of data that will reveal the actual incidence of disconnection failures for those "connectionless" connections.

CHAPTER 11

Quality of Voice for Interactive Multimedia Exchange

All the discussions of quality of packet-switched voice transmission so far have focused on applications to general-use telephony. However, there remains from the typology of packet-switched voice services presented in Part 1 another application that does not have anything to do with telephony. It was described there as *interactive multimedia exchange,* and defined to be an application in which the voice service complements and enhances other exchanges of information via the packet-switched data network. The difference is that where applications of packet-switched voice transmission in hybrid transport and packet-switched telephone services involve tailoring of packet-switched data transmission to support a different function not currently implemented, the voice transmissions in this context add a feature to functions that are implemented.

As an overlay on an existing packet-switched data network, interactive multimedia exchange will support creation of new capabilities in the host medium. For example, VoIP in this application might enable creation of such attractive new Internet facilities as web-shopping where the user is assisted by live dialog with a salesperson who would answer questions as the user browses a web-hosted catalog, IP-hosted videoconferencing, picture telephones implemented on PCs, and use of a PC as the station set for general telephony.

User Concerns

As was observed earlier in Part 1, the principal benefits realized by users of interactive multimedia exchange will be access to telecommunications capabilities that either do not exist today or are only crudely and ineffectively implemented. Experience shows that when a service supports new capabilities like this, for which there are no existing comparable capabilities, users tend to be content to have the capability in any form, as long as minimal quality requirements are met.

For this reason, user concerns with voice quality in interactive multimedia exchanges can be expected to be minimal and expressed as requirements, rather than questions, stating, for example,

- I have to be able to correctly understand what is being said without straining.
- The voice has to sound natural, rather than like something being generated by a robot, so that I can at least think that I am talking to another human,

- I should hear responses to what I say, or ask, that are predictable and that don't take so long that I am left wondering whether they are coming. However, I neither expect nor need those responses to be as rapid as they are when I am arguing with someone over the telephone.

Implications

To the extent that this postulated profile of concerns with interactive multimedia exchanges is accurate, it can be surmised that evaluation of this kind of packet-switched voice service will involve two kinds of differences from evaluation of telephony:

- *Different criteria for acceptability.* The measures of quality of voice connections described for telephony such as a MOS will reflect the effects that inhibit satisfaction of the first two requirements in the previous list. However, in view of the likely user expectations for such a service, values of the MOS that would be evaluated as being indicative of unacceptable quality for telephony may, in fact, represent completely acceptable service in an interactive multimedia exchange application.
- *Different models for effects.* The third requirement expressed in the previous list suggests another aspect of differences in measurement and evaluation of interactive multimedia exchange services. Translated into the language of quantitative evaluation of quality, that statement asserts, in effect, that the marginal effect of round-trip delay on P[UDI] will be less for these services than it is for telephone services, or equivalently, the I_d factor as a function on round-trip delay used in the E-model should be lower.

In addition, the envisioned nature of interactive multimedia exchanges may create dimensions of quality that are not reflected in the measures used for telephony. For example, one likely use of interactive multimedia exchanges will be creation of a picture phone on a personal computer, using the screen to display the video, and the sound card and interactive multimedia exchange for the conversational exchanges. In such an application, one of the possible impairments to user perception of quality will be poor synchronization between the received video and the audio. In fact, a persistent out-of-synch condition will rapidly be rejected by users as "disconcerting" and "unacceptable."

Such possibilities imply that even though extant multiple-effects models might be adjusted for inherent differences between interactive multimedia exchange and telephone services, it is probably more prudent to develop a separate technology for this application. Such a development program would ideally be initiated with the production of a subjective test protocol for interactive multimedia exchange based on education of test subjects to recognize and report the impairments comprising both those defined for telephony and others whose occurrence or impact is unique to interactive multimedia exchanges. Given the data from subjective tests of these kinds, it would then be a short step to extending the VQES, surely, and the E-model, possibly, to this class of applications.

AFTERWORD

There will undoubtedly be some who reach this point with a strong desire to ask, "Why are you telling me all of this? You are belaboring perspectives, concepts, and general functions at the expense of the details of calculation of measures and their analysis that I could use for my immediate problems." The reason, quite simply put, is that foundations are timeless, while the half-life of the whiz-bang, endorsed by everybody, just-gotta-use technique for measurement of voice quality in packet-switched services seems to be about 2 to 3 years. If I took the time in this book to tell you exactly how to implement PAMS, for example, I would not only be treading on intellectual property rights, I would be belaboring something of historical, rather than current, interest now that there is a PESQ. Similarly, the E-model I could detail for you today is sure to be supplanted in the near future with codec-dependent Ie values like those generated with VQmon. Moreover, a lot of detail on many of the subjective tests that have been conducted for the packet-switched environment has been thrown into doubt by the relatively recent recognition of the inherent "burstiness" of dropped packets.

The point, then, is this: When you are confronted with the problem of selecting from new offerings that are sure to come along those that will be cost-effective for the pressing problem of the day, the perspectives, concepts, and general functions presented here are likely to be the only tools that you can rely on to help you.

APPENDIX A

TYPES OF CODECS

One of the technical terms bandied about freely in this book is *codec*. As described in Part 1, the term is shorthand for "coder/decoder," wherein it is understood that what is encoded is an injected acoustic waveform and what is produced by decoding is an image of that acoustic waveform.

By technical description and nomenclature there are something like a dozen different standard codecs for use in telephony, usually distinguished by a leading *G* in the nomenclature. There are, in addition, at least eight technical abbreviations, like PCM (pulse-code modulation), ADPCM (adaptive delta-modulation pulse-code modulation), and CELP (code-excited linear predictive) coding describing different versions of processes used to encode waveforms.

For our purposes, however, it is sufficient to know that there are but two basic types of codecs, depending on whether the encoding produces *digital images* or *functional descriptions* of the injected waveform. The differences are described here.

Digital Images

A digital image of a continuous electronic waveform is a collection of values of the amplitude of the waveform sampled at regular intervals. For nearly all common telephone codecs, the sampling rate is 8000 points per second, producing a sampling interval of 0.125 ms (1/8000 of a second). At each of those sample intervals, the amplitude of the waveform is measured and recorded. The recorded values are then transcribed into digital representations that carry the most significant digits of the recorded value that can be unambiguously mapped into a fixed number of bits.

Notice that we are talking here about a mapping, rather than the simple binary representation of the recorded value, because it is necessary to maintain a constant data transmission rate for the digital signal carrying the encoded value to the sampled amplitude. To accomplish this without relying on strings of bits that are too long to be practicable, a technique called *companding* was developed, whereby integer values that would be exactly represented by 12 bits were mapped onto only 8 bits. One of those mappings is shown in Table A-1. (It is included here, as a gratuitous gift, because few people have ever actually seen the 8-bit μ-law companding map in its simplest form.) According to that scheme, for example, an amplitude measured to be −7904 units would be mapped into a byte comprising all 0s which would be decoded at the far end as an amplitude value of −8031, which would be within 2 percent of the value measured. Similarly, a value of −464 units would be mapped into 01000000, which would be decoded at the far end as −471, to achieve the same accuracy as the larger number.

The digital images received at the distant codec are thus decoded byte by byte to values of pulses of electric energy to put onto the outgoing analog line. This method of generating the electroacoustic waveform at the distant end, which is actually the decoding process, gives the technique its name—pulse-code modulation.

Functional Descriptions

Creation and transmission of digital images of waveforms is a very good technique for transmitting voice over digital links. At 64,000 bits/s, 8-bit companded PCM generates electroacoustic waveforms that are to the ear almost indistinguishable from what was injected. However, PCM-based codecs operating at lower data rates of 32,000 and 16,000 bit/s do not do as well in preserving the original waveform.

Codecs that overcome this limitation have been developed, but they are based on processing that is predicated on the assumption that the electroacoustic waveforms that are to be transmitted are human speech. It works like this:

- Because of the physics of vocal production, there are a very limited number of shapes that small segments of voice can take.
- Because of the physiology of hearing, much of the variation in those shapes is filtered out in the process via which the ear receives aural signals and the nervous system converts them into recognized phonemes and vocal qualities (see the very short course in the physiology of speech and hearing in Chap. 7).
- Consequently, instead of mapping amplitudes point by point to create digital images of speech signals, it should be possible to map

TABLE A-1

Mapping of Sampled Amplitudes onto 8 Bits

0	00000000	-8031	64	01000000	-471	128	10000000	8031	192	11000000	471
1	00000001	-7775	65	01000001	-455	129	10000001	7775	193	11000001	455
2	00000010	-7519	66	01000010	-439	130	10000010	7519	194	11000010	439
3	00000011	-7263	67	01000011	-423	131	10000011	7263	195	11000011	423
4	00000100	-7007	68	01000100	-407	132	10000100	7007	196	11000100	407
5	00000101	-6751	69	01000101	-391	133	10000101	6751	197	11000101	391
6	00000110	-6495	70	01000110	-375	134	10000110	6495	198	11000110	375
7	00000111	-6239	71	01000111	-359	135	10000111	6239	199	11000111	359
8	00001000	-5983	72	01001000	-343	136	10001000	5983	200	11001000	343
9	00001001	-5727	73	01001001	-327	137	10001001	5727	201	11001001	327
10	00001010	-5471	74	01001010	-311	138	10001010	5471	202	11001010	311
11	00001011	-5215	75	01001011	-295	139	10001011	5215	203	11001011	295
12	00001100	-4959	76	01001100	-279	140	10001100	4959	204	11001100	279
13	00001101	-4703	77	01001101	-263	141	10001101	4703	205	11001101	263
14	00001110	-4447	78	01001110	-247	142	10001110	4447	206	11001110	247
15	00001111	-4191	79	01001111	-231	143	10001111	4191	207	11001111	231
16	00010000	-3999	80	01010000	-219	144	10010000	3999	208	11010000	219
17	00010001	-3871	81	01010001	-211	145	10010001	3871	209	11010001	211
18	00010010	-3743	82	01010010	-203	146	10010010	3743	210	11010010	203
19	00010011	-3615	83	01010011	-195	147	10010011	3615	211	11010011	195
20	00010100	-3487	84	01010100	-187	148	10010100	3487	212	11010100	187
21	00010101	-3359	85	01010101	-179	149	10010101	3359	213	11010101	179
22	00010110	-3231	86	01010110	-171	150	10010110	3231	214	11010110	171
23	00010111	-3103	87	01010111	-163	151	10010111	3103	215	11010111	163
24	00011000	-2975	88	01011000	-155	152	10011000	2975	216	11011000	155
25	00011001	-2847	89	01011001	-147	153	10011001	2847	217	11011001	147
26	00011010	-2719	90	01011010	-139	154	10011010	2719	218	11011010	139
27	00011011	-2591	91	01011011	-131	155	10011011	2591	219	11011011	131
28	00011100	-2463	92	01011100	-123	156	10011100	2463	220	11011100	123
29	00011101	-2335	93	01011101	-115	157	10011101	2335	221	11011101	115
30	00011110	-2207	94	01011110	-107	158	10011110	2207	222	11011110	107
31	00011111	-2079	95	01011111	-99	159	10011111	2079	223	11011111	99
32	00100000	-1983	96	01100000	-93	160	10100000	1983	224	11100000	93
33	00100001	-1919	97	01100001	-89	161	10100001	1919	225	11100001	89
34	00100010	-1855	98	01100010	-85	162	10100010	1855	226	11100010	85
35	00100011	-1791	99	01100011	-81	163	10100011	1791	227	11100011	81
36	00100100	-1727	100	01100100	-77	164	10100100	1727	228	11100100	77
37	00100101	-1663	101	01100101	-73	165	10100101	1663	229	11100101	73
38	00100110	-1599	102	01100110	-69	166	10100110	1599	230	11100110	69
39	00100111	-1535	103	01100111	-65	167	10100111	1535	231	11100111	65
40	00101000	-1471	104	01101000	-61	168	10101000	1471	232	11101000	61
41	00101001	-1407	105	01101001	-57	169	10101001	1407	233	11101001	57
42	00101010	-1343	106	01101010	-53	170	10101010	1343	234	11101010	53
43	00101011	-1279	107	01101011	-49	171	10101011	1279	235	11101011	49
44	00101100	-1215	108	01101100	-45	172	10101100	1215	236	11101100	45
45	00101101	-1151	109	01101101	-41	173	10101101	1151	237	11101101	41
46	00101110	-1087	110	01101110	-37	174	10101110	1087	238	11101110	37
47	00101111	-1023	111	01101111	-33	175	10101111	1023	239	11101111	33
48	00110000	-975	112	01110000	-30	176	10110000	975	240	11110000	30
49	00110001	-943	113	01110001	-28	177	10110001	943	241	11110001	28
50	00110010	-911	114	01110010	-26	178	10110010	911	242	11110010	26
51	00110011	-879	115	01110011	-24	179	10110011	879	243	11110011	24
52	00110100	-847	116	01110100	-22	180	10110100	847	244	11110100	22
53	00110101	-815	117	01110101	-20	181	10110101	815	245	11110101	20
54	00110110	-783	118	01110110	-18	182	10110110	783	246	11110110	18
55	00110111	-751	119	01110111	-16	183	10110111	751	247	11110111	16
56	00111000	-719	120	01111000	-14	184	10111000	719	248	11111000	14
57	00111001	-687	121	01111001	-12	185	10111001	687	249	11111001	12
58	00111010	-655	122	01111010	-10	186	10111010	655	250	11111010	10
59	00111011	-623	123	01111011	-8	187	10111011	623	251	11111011	8
60	00111100	-591	124	01111100	-6	188	10111100	591	252	11111100	6
61	00111101	-559	125	01111101	-4	189	10111101	559	253	11111101	4
62	00111110	-527	126	01111110	-2	190	10111110	527	254	11111110	2
63	00111111	-495	127	01111111	0	191	10111111	495	255	11111111	0

collections of amplitudes into a basic waveform shape and some auxiliary values that determine the exact features of that shape.

This, in a nutshell, then, is how codecs based on functional descriptions of expected speech waveforms are implemented. What happens is that instead of a single amplitude, a segment of the incoming electroacoustics waveform is sampled and processed to determine:

- Which of a selection of basic speech waveform patterns it most closely resembles
- What parameters for generating that shape best fit the observed amplitude values

In practice, the size of the segment analyzed is from 10 to 30 ms, depending on the design of the codec. To fix ideas, suppose we are dealing with a 10-ms sample. Then this represents a collection of 80 bytes of amplitude data. With the G.729 codec that is based on 10-ms samples, the 80 bytes of data are reduced to the 10 bytes of data shown in Table A-2. The values for the parameters shown there are selected by statistical analysis to represent the most accurate functional description of the continuous 10-ms segment of voice waveform sampled. At the far end these are decoded to generate either a digital waveform for further transmission or an electroacoustic waveform segment that represents the best approximation to what was injected.

Because the compactness of the functional description of 10 ms of speech is achieved in this way, the G.729, so-called CELP, codec reduces the data rate required for transmission of the digitized voice signal from

TABLE A-2

Functional Description of a 10-ms Sample of a Speech Waveform Transmitted by the G.729 CELP Codec

Code Word Allocation	Number of Bits
Line spectrum pairs	18
Adaptive codebook delay	13
Pitch delay parity	1
Fixed codebook index	26
Fixed codebook sign	8
Codebook gains	6
Codebook gains	8
Total:	80 bits = 10 bytes

64,000 to 8,000 bits/s. Because of the experience with lower-data-rate PCM codecs there is a tendency to assume that the voice signal reconstructed with the CELP codec will have much lower fidelity than the 64,000-bit/s PCM codec. However, subjective tests have consistently shown that absent the differential effects of dropped frame rates the 8000-bit/s CELP codec delivers a voice signal that is perceived by users to be as good as or better than that delivered with a standard 64,000-bit/s PCM codec. Upon a little reflection this should not be a surprise. If the basic premises that support CELP codecs are correct, the curve fitting achieved in the CELP encoding really is, in essence, cleaning up the injected voice signal. The functional representation of the "pure" speech signal received automatically filters out any sporadic noise, low-level noise, and random bit errors that were added to the injected speech signal in the course of its transmission to the codec.

APPENDIX B

HOW A JITTER BUFFER WORKS

Although there is a buffer involved, the name *jitter buffer* for the provisions made to mitigate the effects of jitter on packet-switched voice transmissions is somewhat misleading. Where the name tends to evoke an image of something like the soak buffers employed with streaming audio or video, the jitter buffer sitting in front of packet-fed voice codecs might more properly be called a packet synchronizer. There is a distinction, however.

For streaming audio or video it is possible that the speed at which the signal is received is significantly slower than the data rate for the playback. Because of this, it is necessary to accumulate and store a fairly long segment of the received signal, to give the transmission a head start on the faster playback. If enough of a segment is captured and stored, then the whole transmission is saved before the playback gets to the end. If not, and nearly every Internet user has experienced this at some time, the playback pauses, and there is a break until an adequate segment is again accumulated. In this scheme of things the buffer is actually a segment of computer memory allocated for the whole file which is written to in background while the playback runs in the foreground.

In the case of the jitter buffer, the buffer is a block of memory large enough to hold some multiple, say N, of the size of the payload of the packets being received. Segments of that block of memory equal to the size of a payload are identified by two dynamic pointers:

- A next-to-be-processed (NTP) pointer, which points to the start of the next segment to be fed into the codec for decoding
- A next-segment (NS[s,i]) pointer that calculates the start point for writing the payload packet with the ith sequence number, $i = 1, N - 1$, after the packet sequence number of the packet next to be processed.

Instead of being fed immediately into the codec for decoding, the payload of the first packet of a voice signal received is stored in the segment of memory identified by the NTP pointer, and a clock is started. After a specified holding time T_D, the first packet is extracted and fed to the codec, and the NTP pointer moves to the next segment of memory.

During the holding time, and while the segment of the signal encoded in the first packet is being transmitted, other packets received are checked for sequence number and written into their proper position in the allocated memory segment as directed by the next segment pointer. The writing and extraction process then runs on continuously, with the pointers running cyclically from 1 to N and back to 1. The result is that any packets that are out of sequence because they arrived earlier than the next in sequence are placed in their proper order. More importantly, any packets whose latency is less than T_D more than the first packet received will be in their proper place at the time needed to maintain an uninterrupted flow of decoded speech signals.

This structure, then, supports three important features of a jitter buffer that might not be recognized just from the name and the basic description of its function:

1. *Ready adaptability.* Because of the structure, the amount of resiliency to jitter afforded by the jitter buffer depends not on the allocated memory, but on the value of T_D. This means that adjustments to accommodate actual jitter being experienced can be made by resetting T_D in periods when no speech is being transmitted to increase or decrease the amount of delay that can be tolerated without resulting in a dropped packet. The only limitation is that T_D cannot be set any greater than $(N - 1) \cdot T_P$, where T_P is the time it takes for the codec to generate and transmit the decoded payload.

2. *Fixed memory allocation.* The dependency of changes in the effect of the jitter buffer on T_D, rather than the amount of data being transmitted, means that an adaptive jitter buffer can be created without adapting the buffer (changing the memory allocation).

3. *Small memory allocation.* Because of the dynamic reuse of memory a jitter buffer requires relatively little allocated memory. In fact, it can be argued that values greater than $N = 2$ or 3 do not buy much improvement, because the addition to the round-trip delay becomes the dominant concern for values of T_D that might be accommodated for larger values of N.

APPENDIX C

EXTENSIONS OF THE LOSS/NOISE GRADE OF SERVICE MODEL

The purpose of this appendix is to illustrate the notion of extensibility of a model by showing how the anchor points in a loss/noise grade of service model enable construction of different predictors of the probability that a call will be rated "good" or better (PGoB). To fix ideas, suppose that subjective testing has been conducted to determine the PGoB for service conforming to the anchor points with the results displayed in Table C-1. Since the characteristics of loss and noise for the two tests were identical, it can only be surmised that there is a difference in the perception of the two groups of test subjects. For the sake of argument, suppose, moreover, that the results for test 1 shown in the table were from tests conducted in 1973, and had been in use as the standard for estimating PGoB from the transmission rating factor for 20 years. Suppose further that the data from test 2 were collected from a group of typical users in 1996 and are felt to more realistically reflect the perceptions of users acclimated to the quality of modern digital telephony.

Then, the problem is to take the old standard equation relating R and PGoB_1, the estimates for the 1973 environment, and produce an equation for estimating PGoB_2, from the newer test results. The old equation from standards books is

$$\text{PGoB}_1 = N_{\text{CUM}}\left[\frac{R - 63.14}{15.33}\right] \qquad \textbf{[C-1]}$$

where N_{CUM} denotes the cumulative distribution of the standard normal distribution with mean 0 and variance 1 (denoted $N[0, 1]$). Selected values for N_{CUM} are shown in Table C-2.

TABLE C-1

Hypothetical Test Results at Anchor Points for the Loss/Noise Grade of Service Model

Transmission Signal Loss, db	Circuit Noise, dBrnC	Transmission Rating Factor R	PGoB Test 1, %	PGoB Test 2, %
15	25	80	86.4	63.9
30	40	40	6.5	1.5

TABLE C-2

Selected Values* of the Cumulative Distribution Function for $N[0, 1]$

$X =$	$N_{CUM} =$
−2.17	0.015
−2.06	0.02
−1.64	0.05
−1.51	0.065
−1.29	0.10
−0.84	0.20
0	0.50
0.26	0.60
0.36	0.639
0.53	0.70
0.84	0.80
1.10	0.864
1.29	0.90

To understand how to build the new equation for estimating PGoB from the test 2 results, it is instructive to understand how the magic equation [C-1] was derived way back in the recesses of time. To help with this, Table C-3 summarizes all the relevant information from Tables C-1 and C-2.

The reason for selecting the data shown in Table C-3 is that Eq. [C-1] was derived by defining PGoB to be N_{CUM} for the normally distributed random variable R. To do this, values of R had to be normalized by finding a mean μ and variance σ such that

$$\frac{R - \mu}{\sigma}$$

is distributed $N[0, 1]$.

The quickest way to determine μ and σ is to note that in a normal distribution the value of μ occurs where $X = 0$ and $N_{CUM} = 0.50$ (the midpoint of the distribution). Setting

*These values were taken from the table "Normal Distribution and Related Functions" on pp. 23 to 33 of the 1966 edition of the *CRC Handbook of Tables for Probability and Statistics,* published by the Chemical Rubber Company, Cleveland, Ohio. Similar tables can be found in almost any book of mathematical tables.

TABLE C-3

Data Used to Produce Eq. [C-1]

R	PGoB	Value of X for N_{CUM} = PGoB
40	.065	−1.51
80	.864	1.10

TABLE C-4

Data Used to Produce the Equation for the Test 2 Environment

R	PGoB	Value of X for N_{CUM} = PGoB
40	.015	−2.17
80	.639	0.36

$$X = R \cdot A + B \qquad \text{[C-2]}$$

and using the values of R and N_{CUM} in Table C-3, create the equations

$$-1.51 = 40A + B \qquad \text{[C-3a]}$$

and

$$1.10 = 80A + B \qquad \text{[C-3b]}$$

from which $A = 2.61/40$ and $B = -4.12$, and the value of R corresponding to $X = 0$ is found to be $[(4.12) \cdot 40]/2.61 = 63.14 = \mu$.

Once μ is found, then σ can be found by setting $\sigma = (X - \mu)/N_{CUM}$ for one of the sets of values in Table C-3. Using $R = 80$, this produces $\sigma = (80 - 63.14)/1.1 = 15.33$.

Given the knowledge of how the original standard Eq. [C-1] was derived, the extension of the loss/noise grade of service model to the test 2 environment is now clear and easily implemented. The necessary data are given in Table C-4. Analogous application of Eqs. [C-2] and [C-3] to that data quickly yields the new, improved, much more credible relationship based on test 2:

$$\text{PGoB}_2 = N_{CUM}\left[\frac{R - 74.31}{15.81}\right] \qquad \text{[C-4]}$$

APPENDIX D

U.S. PATENTS FOR VOICE QUALITY MEASUREMENT AND ANALYSIS TECHNOLOGY

This appendix contains identification and extracts from some of the U.S. patents that have been issued for various voice quality measurement and analysis techniques. Each extract identifies

- Patent number and title
- Date of the application and issue, to convey some sense of the progress of the technology
- Names of the inventors and the assignee
- Abstract for the patent
- Citations, comprising references to open literature and earlier patents

In addition, there may be extracts from the body of the patent. These have been selected to illuminate some particular aspects of the invention, such as the motivation for the development, the perception of the problem and its solution, or its design, that are particularly interesting in the context of this book. Commentary added to facilitate apprehension of the important material is provided as footnotes. Except for such commentary and formatting, and corrections of obvious typographical errors in references cited, all other material included for any patent is directly quoted, with only the most minor changes for clarity shown only within square brackets, from the patent listings found on the Internet at *www.uspto.gov/main/patents.htm.*

For ease of reference, the patents have been divided into six groups:

- Early patents of historical interest
- Multitone test techniques
- PAMS technology
- Extensions of PAMS technology
- PSQM technology
- Technology employed in the VQES

Interested readers are encouraged to go to the original source materials available on the U.S. Patent and Trademark Office website.

Early Patents of Historical Interest

Patent No. 4,152,555

Title: Artificial ear for telephonometric measurements

Date Issued: May 1, 1979

Date of Application: March 8, 1978

Inventors: Modena; Giulio (Caselle, IT); Reolon; Aldo (Chieri-Torino, IT)

Assignee: CSELT—Centro Studi e Laboratori Telecomunicazioni S.p.A. (Turin, IT)

Abstract

A device to be used in phonometric measurements on telephone equipment, simulating a human ear, has a generally disk-shaped body with an annular ridge encompassing a frustoconical central recess serving as a wide-open entrance cavity. The body has several internal cavities communicating with the recess through restricted channels, namely a pair of major cavities resonant at low audio frequencies, an intermediate cavity resonant in the middle audio range, and a minor cavity resonant at high audio frequencies, the last-mentioned cavity having the shape of a narrow cylinder extending generally along the body axis. A microphone, connected to a measuring circuit, rises from the bottom of the recess to substantially the top level of the surrounding ridge to pick up incoming sounds at that level.

Citations

U.S. Patent Documents

3744294; Jul. 1973; Lewis et al.; Acoustical transducer calibrating system and apparatus

Other References

Green Book, vol. V, Recommendation p. 51, "...Artificial Ear...," recommendation by CCITT of the IEC.

Funk-Technik, 32, Jahrgang, Nr. 5/1977, "A New Coupling Device for Headphone Measurement."

Extracts

Background of the Invention So-called telephonometric measurements, designed to test the performance of electroacoustic transducers such as the receivers and transmitters of telephone handsets, are advantageously carried out automatically with the aid of devices simulating human ears and mouths. This not only saves manpower but also allows the standardization of testing equipment according to internationally established specifications.

Thus, an artificial ear of the type here envisaged is a phonometric device which acts as an acoustic load for a telephone receiver and whose sensitivity/frequency characteristic should correspond as closely as possible to that of the human ear. A microphone forming part of the device translates the incoming sound waves into electrical signals which are sent to a measuring circuit for evaluation of the response characteristic of the receiver undergoing testing.

The International Electrotechnical Commissioner (IEC) has proposed an artificial ear whose adoption for telephonometric measurements was provisionally recommended by the CCITT during its 5th Plenary Assembly (see Green Book, Vol. V, recommendation P51).

The IEC artificial ear simulates the performance of a human ear whose auricle or pinna is tightly pressed against the earpiece of a telephone handset so that no acoustic leakages occur between the telephone receiver and the ear. In practice, however, a user will press the receiver tightly against his ear only under extraordinary circumstances, as where the signal is very faint or the telephone is located in a noisy room. Normally, the handset is held close to the ear but with enough clearance to generate significant acoustic leakage.

Thus, a telephone receiver tested with the IEC artificial ear and found to have a substantially frequency-independent response may not perform satisfactorily in actual use.

Object of the Invention The object of our present invention, therefore, is to provide an improved phonometric device for the purpose set forth which more faithfully reproduces the conditions of sound reception by a human ear held close to a telephone receiver.

Summary of the Invention A phonometric device according to our invention comprises a generally disk-shaped body with an annular ridge encompassing a substantially frustoconical, outwardly diverging recess which may be termed an entrance cavity for sound waves emanating from a telephone receiver placed on that ridge, the sound waves being converted into electrical signals by a microphone disposed in the recess. The body is provided with several internal cavities communicating with the recess, namely one or preferably two major cavities resonant at a low audio frequency, an intermediate cavity resonant in the middle audio range, and a minor cavity resonant at a high audio frequency. The minor cavity is elongate, preferably cylindrical, and opens onto the recess near the bottom thereof.

According to a more particular feature of our invention, the major and intermediate cavities communicate with the recess through restricted channels which are of acoustically resistive and inductive character while the cavities themselves are essentially capacitive. The minor cavity, which preferably extends substantially along the axis of the body and its frustoconical recess, acts as an acoustical transmission line with distributed constants.

Pursuant to a further feature of our invention, both the intermediate and minor cavities are separated from the recess or entrance cavity by substantially pure acoustic resistances.

The depth of the recess, generally on the order of 1 cm, corresponds to about one acoustic wavelength at a frequency between 3 and 4 kHz. In order to minimize the effect of the resulting phase delay, we prefer to dispose the pick-up head of the microphone substantially at the level of the crest of the ridge, i.e. at the broad base of the frustoconical recess, in the immediate vicinity of the receiver to be tested.

Patent No. 4,860,360*

Title: Method of evaluating speech

Date Issued: August 22, 1989

Date of Application: April 6, 1987

Inventors: Boggs; George J. (Weston, MA)

Assignee: GTE Laboratories Incorporated (Waltham, MA)

Abstract

A method of evaluating the quality of speech in a voice communication system is used in a speech processor. A digital file of undistorted speech representative of a speech standard for a voice communication system is recorded. A sample file of possibly distorted speech carried by said voice communication system is also recorded. The file of standard speech and the file of possibly distorted speech are passed through a set of critical band filters to provide power spectra which include distorted-standard speech pairs. A variance-covariance matrix is calculated from said pairs, and a Mahalanobis D.sup.2 calculation is performed on said matrix, yielding D.sup.2 data which represents an estimation of the quality of speech in the sample file.

Citations

U.S. Patent Documents

3634759; Jan., 1972; Tokorozawa et al.; Frequency spectrum analyzer with a real time display device

4220819; Sep., 1980; Atal; Residual excited predictive speech coding system

4509133; Apr., 1985; Monbaron et al.; Apparatus for introducing control words by speech

4592085; May, 1986; Watari et al.; Speech-recognition method and apparatus for recognizing phonemes in a voice signal

4651289; Mar., 1987; Maeda et al.; Pattern recognition apparatus and method for making same

*This is the most frequently cited prior art patent in patents for measurement of speech distortion.

Other References

Klatt, "A Digital Filter Bank for Spectral Matching," IEEE ICASSP, 1976, pp. 573–576.

Campbell et al., "Voiced/Unvoiced Classification of Speech with Applications to the U.S. Government LPC-10E Algorithm," ICASSP 86, Tokyo, pp. 473–476, 1986.

Extracts

Background of the Invention Speech quality judgments in the past were determined in various ways. Subjective, speech quality estimation was made by surveys conducted with human respondents. Some investigators attempted to evaluate speech quality objectively by using a variety of spectral distance measures, noise measurements, and parametric distance measures. Both the subjective techniques and the prior objective techniques were widely used, but each has its own unique set of disadvantages.

The purpose of speech quality estimation is to predict listener satisfaction. Hence, speech quality estimation obtained through the use of human respondents (subjective speech quality estimates) is the procedure of choice when other factors permit. Disadvantageously, the problems with conducting subjective speech quality studies often either preclude speech quality assessment or dilute the interpretation and generalization of the results of such studies.

First and foremost, subjective speech quality estimation is an expensive procedure due to the professional time and effort required to conduct subjective studies. Subjective studies require careful planning and design prior to the execution. They require supervision during execution and sophisticated statistical analyses are often needed to properly interpret the data. In addition to the cost of professional time, human respondents require recruitment and pay for the time they spend in the study. Such costs can mount very quickly and are often perceived as exceeding the value of speech quality assessment.

Due to the expense of the human costs involved in subjective speech quality assessment, subjective estimates have often been obtained in studies that have compromised statistical and scientific rigor in an effort to reduce such costs. Procedural compromises invoked in the name of cost have seriously diluted the quality of the data with regard to their generalization and interpretation. When subjective estimates are not generalized beyond the sample of people recruited to participate in the study, or even when the estimates are not generalized beyond some subpopulation within the larger population of interest, the estimation study has little real value. Similarly, when cost priorities result in a study that is incomplete from a statistical perspective (due to inadequate controlled conditions, unbalanced listening conditions, etc.), the interpretation of the results may be misleading. Disadvantageously, inadequately designed studies have been used on many occasions to guide decisions about the value of speech transmission techniques and signal processing systems.

Because cost and statistical factors are so common in subjective speech quality estimates, some investigators have searched for objective methods to replace the subjective methods. If a process could be developed that did not require human listeners as speech quality judges, that process would be of substantial utility to the voice communication industry and the professional speech community. Such a process would enable speech scientists, engineers, and product customers to quickly evaluate the utility of speech systems and quality of voice communication systems with minimal cost. There have been a number of efforts directed at designing an objective speech quality assessment process.

The prior processes that have been investigated have serious deficiencies. For example, an objective speech quality assessment process should correlate well with subjective estimates of speech quality and ideally achieve high correlations across many different types of speech distortions. The primary purpose for estimating speech quality is to predict listener satisfaction with some population of potential listeners. Assuming that subjective measures of speech quality correlate well with population satisfaction (and they should, if assessment is conducted properly), objective measures that correlate well with subjective estimates will also correlate well with population satisfaction levels. Further, it is often true that any real speech processing or voice transmission system introduces a variety of distortion types. Unless the objective speech quality process can correlate well with subjective estimates across a variety of distortion types, the utility of the process will be limited. No objective speech quality process previously reported in the professional literature correlated well with subjective measures. The best correlations obtained were for [a] limited set of distortions.

Summary of the Invention It is the principal object of this invention to provide for a new and improved objective process for evaluating speech quality by incorporating models of human auditory processing and subjective judgment derived from psychoacoustic research literature.

Another object of this invention is to provide for a new and improved objective process of evaluating the quality of speech that correlates well with subjective estimates of speech quality, wherein said process can be over a wide set of distortion types.

Yet another object of this invention is to provide for a new and improved objective method of evaluating speech quality that utilizes software and digital speech data.

Still another object of this invention is to provide for a new and improved objective method of evaluating speech quality in which labor savings for both professional and listener time can be substantial.

In accordance with one aspect of this invention, a method of evaluating the quality of speech through an automatic testing system includes a plurality of steps. They include the preparation of input files. The first type of input file is a digital file of undistorted or standard speech utilizing a human voice. A second type of input file is a digital file of distorted speech. The standard speech [is] passed through the system to provide at least one possibly somewhat distorted speech file, since at least one distorted speech file is necessary to use the invention. A set of critical band filters is selected to encompass the bandpass characteristics of a communications network. The standard speech and the possibly distorted speech are passed through the set of filters to provide power spectra relative thereto. The power spectra obtained from the standard speech file and from the possibly somewhat distorted speech file are temporarily stored to provide a set of distorted-standard speech pairs. A variance-covariance matrix is prepared from the set of distorted-standard speech pairs....

In accordance with certain features of the invention, the standard speech is prepared by digitally recording a human voice on a storage medium, and the set of critical band filters is selected to encompass the bandpass characteristics of the international telephone network (nominally 300 Hz to 3200 Hz). The set of filters can include fifteen filters having center frequencies, cutoff frequencies, and bandwidths, where the center frequencies range from 250 to 3400 Hz, the cutoff frequencies range from 300 to 3700 Hz, and the bandwidths range from 100 to 550 Hz. The center frequency is defined as that frequency in which there is the least filter attenuation. In such a method, the set of filters can include sixteen filters, the sixteenth filter having a center frequency of 4000 Hz, a cutoff frequency of 4400 Hz, and a bandwidth of 700 Hz.... The possibly somewhat distorted speech can be recorded by various means including digital recording. The spectra from the standard speech and the possibly somewhat distorted speech file from the set of critical band filters can be temporarily stored via parallel paths. It can be temporarily stored by a serial path.

Patent No. 5,369,644

Title: Method for testing a communication system using canned speech

Date Issued: November 29, 1994

Date of Application: December 16, 1991

Inventors: Oliver; Douglas W. (Watauga, TX)

Assignee: Motorola, Inc. (Schaumburg, IL)

Abstract

A system is tested by providing a canned speed signal (a signal containing known speech) from a diagnostic controller to an encoder of the communication system. The canned signal is encoded forming an encoded canned speech signal. The encoded signal is transmitted to an where it is then decoded. The decoded signal is compared with the original canned speech signal in the diagnostic controller. A report of the comparison is then provided to a maintenance facility.

Citations

U.S. Patent Documents

Re30037; Jun., 1979; Bass; Data communications network remote test and control system

4744083; May, 1988; O'Neil et al.; Satellite-based position determining and message transfer system with monitoring of link quality

4831624; May, 1989; McLaughlin et al.; Error detection method for subband coding

5206864; Apr., 1993; McConnell; Speech synthesis apparatus and method

Extracts

Background of the Invention A continuing problem in the field of communications is the ability to test the systems in order to improve their maintainability. In current testing, one or more of the elements of the communication system is replaced with a substitute testing device. While this permits portions of the system to be tested, the testing device eliminates an important element, such as the decoder, from the system. Since errors can result from the interaction of various elements of the system, the testing is incomplete unless all of the system elements are present during the testing.

Summary of the Invention A method is described for testing a communication system which utilizes a canned speech signal. The canned speech signal is provided from a diagnostic controller to an encoder of the communication system. The canned signal is encoded forming an encoded canned speech signal. The encoded signal is transmitted to a decoder. The decoded signal is then compared with the original canned speech signal in the diagnostic controller. A report of the comparison is then provided to a maintenance facility.

Patent No. 5,490,234*

Title: Waveform blending technique for text-to-speech system

Date Issued: February 6, 1996

Date of Application: January 21, 1993

Inventors: Narayan; Shankar (Palo Alto, CA)

Assignee: Apple Computer, Inc. (Cupertino, CA)

Abstract

A concatenator for a first digital frame with a second digital frame, such as the ending and beginning of adjacent diphone strings being concatenated to form speech is based on determining an optimum blend point for the first and second digital frames in response to the magnitudes of samples in the first and second digital frames. The frames are then blended to generate a digital sequence representing a concatenation of the first and second frames with reference to the optimum blend point. The system operates by first computing an extended frame in response to the first digital frame, and then finding a subset of the extended frame [which] matches the second digital frame using a minimum average magnitude difference function over the samples in the subset. The blend point is the first sample of the matching subset. To generate the concatenated waveform, the subset of the extended frame is combined with the second digital frame and concatenated with the beginning segments of the extended frame to produce the concatenate waveform.

Citations

U.S. Patent Documents

4384169; May, 1983; Mozer et al.; Method and apparatus for speech synthesizing

4692941; Sep., 1987; Jacks et al.; Real-time text-to-speech conversion system

4852168; Jul., 1989; Sprague; Compression of stored waveforms for artificial speech

---

*This patent is only of passing interest for our purposes. It is included because it evinces the importance of intersyllabic transitions in maintaining natural sounding voice.

5153913; Oct., 1992; Kandefer et al.; Generating speech from digitally stored coarticulated speech segments

5220629; Jun., 1993; Kosaka et al.; Speech synthesis apparatus and method

5327498; Jul., 1994; Hamon; Processing device for speech synthesis by addition overlapping of wave forms

Extracts

Background of the Invention In text-to-speech systems, stored text in a computer is translated to synthesized speech....However in prior art systems which have reasonable cost, the quality of the speech has been relatively poor making it uncomfortable to use or difficult to understand....

In text-to-speech systems, an algorithm reviews an input text string, and translates the words in the text string into a sequence of diphones which must be translated into synthesized speech. Also, text-to-speech systems analyze the text based on word type and context to generate intonation control used for adjusting the duration of the sounds and the pitch of the sounds involved in the speech.

Diphones consist of a unit of speech composed of the transition between one sound, or phoneme, and an adjacent sound, or phoneme. Diphones typically start at the center of one phoneme and end at the center of a neighboring phoneme. This preserves the transition between the sounds relatively well.

American English based text-to-speech systems, depending on the particular implementation, use about fifty different sounds referred to as phones. Of these fifty different sounds, the standard language uses about 1800 diphones out of possible 2500 phone pairs. Thus, a text-to-speech system must be capable of reproducing 1800 diphones. To store the speech data directly for each diphone would involve a huge amount of memory. Thus, compression techniques have evolved to limit the amount of memory required for storing the diphones.

Two concatenated diphones will have an ending frame and a beginning frame. The ending frame of the left diphone must be blended with the beginning frame of the right diphone without audible discontinuities or clicks being generated. Since the right boundary of the first diphone and the left boundary of the second diphone correspond to the same phoneme in most situations, they are expected to be similar looking at the point of concatenation. However, because the two diphone codings are extracted from different contexts, they will not look identical. Thus, blending techniques of the prior art have attempted to blend concatenated waveforms at the end and beginning of left and right frames, respectively. Because the end and beginning of frames may not match well, blending noise results. Continuity of sound between adjacent diphones is thus distorted.

Summary of the Invention The present invention provides an apparatus for concatenating a first digital frame with a second digital frame of quasi-periodic waveforms, such as the ending and beginning of adjacent diphone strings being concatenated to form speech. The system is based on determining an optimum blend point for the first and second digital frames in response to the magnitudes of samples in the first and second digital frames. The frames are then blended to generate a digital sequence representing a concatenation of the first and second frames, with reference to the optimum blend point. This has the effect of providing much better continuity in the blending or concatenation of diphones in text-to-speech systems than has been available in the prior art.

Multitone Test Techniques

Patent No. 4,301,536

Title: Multitone frequency response and envelope delay distortion tests

Date Issued: November 17, 1981

Date of Application: December 28, 1979

Inventors: Favin; David L. (Little Silver, NJ); Lynn; Peter F. (Little Silver, NJ); Snyder; Paul J. (Linden, NJ)

Assignee: Bell Telephone Laboratories, Incorporated (Murray Hill, NJ)

Abstract

Accurate and reliable measurement results are obtained in a test system...employing digital data acquisition units (121) when measuring frequency response or envelope delay distortion of a network or communication facility (105) by employing a unique test signal (21-tone) including a plurality of tones, each tone having amplitude, frequency, phase component values determined and assigned in accordance with prescribed criteria. The phase component values are determined in accordance with a relationship dependent on the number of tones in the test signal, and in one example, are initially assigned on a random, one-to-one basis to the tones. In a specific embodiment, a test signal is utilized having 21 tones. Further problems arising from nonlinearities on the facility under evaluation (105) are minimized by transmitting the 21-tone test signal a plurality of times and by reassigning the phase component values to the tones each time the test signal is transmitted. In a specific example, the phase component values are reassigned to the tones in counterclockwise, circular fashion after each transmission of the test signal until each tone has taken on each phase component value. Once measurements, i.e., data records, of the transmitted test signal have been obtained, they are utilized to obtain a spectrum, and in turn, the frequency components of the spectrum are employed to obtain the desired measurement of frequency response or envelope delay distortion.

Citations

U.S. Patent Documents

3271666; Sep., 1966; Anderson et al.; Apparatus for measuring envelope delay distribution wherein selected impulses of a high frequency standard are gated to one input of a bistable phase comparator

3573611; Apr., 1971; Bergemann et al.; Simultaneous delay measurement between selected signal frequency channel and reference frequency channel

3842247; Oct., 1974; Anderson; Apparatus and method for measuring envelope delay utilizing π-point technique

3970926; Jul., 1976; Rigby et al.; Method and apparatus for measuring the group delay characteristics of a transmission path

4001559; Jan., 1977; Osborne et al.; Programmable measuring

4039769; Aug., 1977; Bradley; Instrument and method for measuring envelope delay in a transmission channel

Extracts

Background of the Invention In order to maintain networks or communication systems properly, for example, telephone transmission facilities and the like, numerous measurements are made of network and system characteristics. Important among these are the measurements of frequency response and envelope delay distortion. To this end, what is commonly called envelope delay is measured over the frequency range of the facility being evaluated. Envelope delay is defined as the slope of the phase versus frequency characteristic of the transmission facility. In an ideal communications system, envelope delay is constant over the frequency band. However, in practical systems there are deviations in the envelope delay over the frequency band. These deviations from an arbitrary reference are defined as the envelope delay distortion of the facility.

Heretofore, envelope delay measurements have been made by employing a carrier frequency signal which is amplitude modulated by a stable "low" frequency reference signal. The carrier frequency and upper and lower sidebands are propagated through the facility being evaluated, thereby experiencing a delay dependent upon their position in the frequency band. These signals are detected at the output of the facility under evaluation. Then, a measure of envelope delay at the carrier frequency is obtained by precisely measuring the delay interval between the detected signals and the low frequency reference signal....

Problems common to the known prior measurement arrangements are errors in and repetition of measurements of envelope delay distortion of facilities on which noise, frequency shift, nonlinearity or other impairments are present. Moreover, it is important to obtain accurate and reliable measurements in the presence of subtle changes in the amount of intermodulation distortion on the facility under evaluation.

Data collected for obtaining envelope delay measurements is also utilized to compute the frequency response of the facility.

Summary of the Invention The problems of accuracy, reproducibility and reliability of measurements and other problems of prior measurement apparatus are overcome in measuring frequency response or envelope delay distortion of a network or facility by employing a set of test signals, each test signal having a plurality of tones and each tone having frequency, amplitude and phase component values determined and assigned thereto in accordance with prescribed criteria. An ensemble of measurements is made while transmitting the set of test signals over the network or facility under evaluation. In turn, the ensemble of measurements is used to obtain values for frequency response and/or envelope delay distortion.

Patent No. 4,417,337

Title: Adaptive multitone transmission parameter test arrangement

Date Issued: November 22, 1983

Date of Application: June 29, 1981

Inventors: Favin; David L. (Little Silver, NJ); Lynn; Peter F. (Little Silver, NJ); Snyder; Paul J. (Linden, NJ)

Assignee: Bell Telephone Laboratories, Incorporated, (Murray Hill, NJ)

Abstract

Accurate and reliable measurements of transmission parameters, e.g., envelope delay distortion or frequency response, of a network or facility (105) are obtained in a test system... employing digital data acquisition units (121) by utilizing a unique test signal including a plurality of tones. A set of test signals is transmitted over the facility (105), a set of measurements is made of the received version for each test signal, each set of measurements is time averaged, and an ensemble of time-averaged sets of measurements is used to generate the desired measurements of the transmission parameters. System efficiency is enhanced by measuring prescribed transmission impairments, e.g., nonlinear distortion (3OID), signal-to-noise ratio (S/N) and frequency shift (FS) on the facility under evaluation and dynamically determining test system parameters in accordance with predetermined relationships with the measured impairments..., namely, the number (D) of test signals in the set of test signals, the maximum number (T) of consecutive measurements to be time-averaged, the required number of measurements (M) and the number of time-averaged measurements to be made in order to obtain a desired measurement accuracy.

Citations

U.S. Patent Documents

3271666; Sep., 1966; Anderson et al.; Apparatus for measuring envelope delay distribution wherein selected impulses of a high frequency standard are gated to one input of a bistable phase comparator

3573611; Apr., 1971; Bergemann et al.; Simultaneous delay measurement between selected signal frequency channel and reference frequency channel

3842247; Oct., 1974; Anderson; Apparatus and method for measuring envelope delay utilizing π-point technique

3970926; Jul., 1976; Rigby et al.; Method and apparatus for measuring the group delay characteristics of a transmission path

4001559; Jan., 1977; Osborne et al.; Programmable measuring

4039769; Aug., 1977; Bradley; Instrument and method for measuring envelope delay in a transmission channel

4264959; Apr., 1981; Blaass; Testing procedure and circuit for the determination of frequency distortions and group delay time distortion of a test object

4275446; Jun., 1981; Blaess; Method and apparatus for measurement of attenuation and distortion by a test object

4,301,536; Nov., 1981; Favin, et al.; Multitone frequency response and envelope delay distortion tests

Extracts

Summary of the Invention More efficient use of system resources is realized in a transmission test arrangement employing a test signal including a plurality of tones by first measuring prescribed impairments of a network or facility under evaluation and, then, dynamically determining in accordance with predetermined relationships with the measured impairments, test system parameters to be employed in making a measurement. Specifically, a dynamic determination is made of the number of test signals in a set to be transmitted over the network or facility, a required number of measurements (data records), the maximum number of consecutive measurements of a received test signal to be time averaged and the number of time-averaged measurements which must be used in order to obtain a desired accuracy in generating measurements of prescribed transmission parameters of the network or facility under evaluation, e.g., frequency response and/or envelope delay distortion.

In a specific embodiment of the invention the number of different test signals required to obtain an accurate measurement is further minimized by obtaining and utilizing a plurality of time-averaged measurements for each test signal. The number of time-averaged measurements is also determined dynamically in accordance with predetermined relationships with the measured impairments.

Patent No. 4,768,203

Title: Signal-to-noise ratio testing in adaptive differential pulse code modulation

Date Issued: August 30, 1988

Date of Application: September 18, 1987

Inventors: Ingle; James F. (Boro of Fair Haven, NJ)

Assignee: Bell Communications Research, Inc. (Livingston, NJ)

Abstract

The adaptive and predictive capabilities of Adaptive Differential Pulse Code Modulation (ADPCM) equipment enable a telecommunication system to maintain acceptable signal/noise levels in voice transmission while utilizing a significantly lower encoding bit rate than that of conventional Pulse Code Modulation (PCM). ADPCM, however, has a deleterious effect on high-speed voiceband data transmission, yet due to its adaptive capabilities cannot readily be identified or evaluated by means of conventional ANSI/IEEE standard test signals and methods. The procedure of the present invention enables such identification and evaluation by imposing upon an ADPCM system a multiple-tone test signal which spans the voiceband and has amplitude characteristics similar to white noise. This signal thereby effectively overloads the adaptive and predictive capabilities of the system and causes the generation of a notably high level of quantizing noise. The resulting multitone signal-with-noise output from the system is processed in a spectrum analyzer where the accumulation of the signal levels in the distinct and narrow input tone bands is compared with the remainder of the accumulated signal power to obtain an accurate signal/noise measurement which, in addition to providing substantive analytical data, yields an indication of the presence of ADPCM, as distinguished even from tandem PCM, encoding equipment in the system.

Citations

U.S. Patent Documents

3639703; Feb., 1972; Bergemann et al.; Method and means for measuring weighted noise in a communication link

3737781; Jun., 1973; Deerkoski; Signal-to-noise ratio determination circuit

4028622; Jun., 1977; Evans et al.; Selective intermodulation distortion measurement

4032716; Jun., 1977; Allen; Measurement of noise in a communication channel

4246655; Jan., 1981; Parker; Circuit for providing an indication of signal/noise ratio in a selector diversity system

4273970; Jun., 1981; Favin et al.; Intermodulation distortion test

4,301,536; Nov., 1981; Favin et al.; Multitone frequency response and envelope delay distortion tests

4355407; Oct., 1982; Mueller; Device for disconnecting the receiver in case of a small signal-to-noise ratio for a digital-modulated radio system

4513426; Apr., 1985; Jayant; Adaptive differential pulse code modulation

Extracts

Background of the Invention Over two-thirds of the telecommunications facilities now in use for toll traffic digitally encode voice and voiceband data by sampling the customer signals at an 8000 Hz rate and encoding each sample into eight digital bits. This process of Pulse Code Modulation (PCM) provides an allocation of 64 kilobits per channel over 24 channels. Added framing information utilizes another 8 kilobits, thus resulting in the common transmission rate of 1.544 megabits per second. Various proposals have been made for increasing the available transmission channels; for example, by decreasing the number of encoding bits for each signal sample and thus substantially lessening the required transmission bandwidth.

The only one of these proposals that has reached significant implementation is the 32 kilobit Adaptive Differential Pulse Code Modulation (ADPCM) system which effectively provides a doubling of capacity to 48 channels at the usual sampling and transmission rates by encoding the customer signal samples into words of four bits each. In this system the values of the four encoding bits are assigned in PCM-to-ADPCM transcoders or in ADPCM digital channel banks by complex algorithms which both predict and adapt to the voiceband signals based upon their magnitude, waveform, frequency content, and frequency spread. The operation of an ADPCM encoding system may be seen in the general description in U.S. Pat. No. 4,513,426.

The relatively limited permutations in the signals derived from human speech make possible the successful operation of ADPCM in that this system can readily predict from previously occurring signal samples the probable magnitude and frequency range of the ensuing signal pattern, and can adapt to the usually moderate variations in these parameters by optimally selecting the encoding bit assignments which will narrowly encompass the pattern range. In this manner an ADPCM system can reasonably reproduce the waveforms associated with speech by means of only four encoding bits.

During telecommunication conversations, listeners are [reasonably] tolerant of noise occurring during speaking intervals if the channel is quiet between the speech bursts, and therefore will not find objectionable the four-bit quantizing noise that is generated in the ADPCM encoding process. The implementation of voiceband data transmission, on the other hand, particularly with high-speed (9.6 kb/s) modem equipment, has created a problem in that these transmissions normally span all but a small fraction of the voice bandwidth and outstrip the capability of the ADPCM system to predict and adapt to their effectively random, wide bandwidth signals. The encoding system is thus unable to sense a trend in the signal to which the adaptive algorithm may be applied, and, therefore, generates simple four-bit encoding of the wide-band signal. The resulting high level of quantizing noise renders the telephone channel incapable of faithfully functioning as a medium for high-speed modem use.

Due to the serious degradation in transmission quality imposed by ADPCM encoding upon systems operating with high-speed modems or with other broad bandwidth applications, standards bodies have imposed restrictions, including complete exclusion, on the number of such systems that may be permitted on an end-to-end connection. Unfortunately, however, since they are based upon the use of test signals which have easily-predictable waveforms, none of the ANSI/IEEE Std. 743-1984 standard voiceband test procedures are capable of identifying the presence of an ADPCM system in a transmission circuit.

Thus, the capabilities of ADPCM encoding systems which enable them to predict and adapt to common voiceband conversation signals allow them to remain largely transparent to identification by previously available, standardized test methods. As a result, such standard test procedures cannot distinguish between a lower signal quality resulting from a single ADPCM system and that resulting from a number of less deleterious PCM systems in tandem, and are ineffective as a means for determining the quality of ADPCM equipment or the contribution such equipment may make to overall noise in a system.

The present invention, on the other hand, provides a method of testing which avoids the predictive and adaptive capabilities of an ADPCM encoding system in order to allow the identification of the presence of such a system in a voiceband telecommunications circuit, and which additionally enables accurate testing of the quality of ADPCM encoding equipment with respect to the generation of spurious noise in a transmission.

Summary of the Invention The present invention provides test equipment and a method for its implementation which impose upon a telecommunication transmission system comprising ADPCM encoding equipment a multiple-tone test signal of near-white-noise waveform which encompasses substantially the entire width of the voiceband, thereby overloading the predictive and adaptive capabilities of the ADPCM system in much the same way as does an operative high-speed modem data transmission signal. Such signal is derived in the manner described in U.S. Pat. No. 4,301,536, and comprises twenty-one tones of narrow bandwidth, or high spectral purity, which extend in uniform frequency distribution across the voiceband between about 200 and 3400 Hz. Another useful test signal may comprise the centrally-disposed fifteen contiguous tones of such a twenty-one tone signal.

The test signal, after transmission through the communications system in which it suffers the perturbations of quantizing noise arising from the ADPCM encoding function, is input to a test circuit comprising an array processor, or spectrum analyzer, which separates the power in the signal into discrete frequency "bins" that include the frequencies of the original test signal tones. The total power in the twenty-one tone bins is measured as the indication of transmitted signal, and the total power in the remaining bins is measured and C-message weighted, according to the ANSI/IEEE standard, to obtain the measure of the noise transmitted with the test signal. These measurements are then used to directly calculate the signal-to-noise ratio (SNR) of the ADPCM system.

Since the bandwidth and phase distribution of the twenty-one tone test signal are such as to cause it to resemble a white-noise signal extending across the voiceband, the adaptation and prediction algorithms of the ADPCM system, which are essential to maintaining an acceptable SNR in the four-bit signal encoding of speech, are unable to be implemented and there results the generation by the system of the expected, larger, four-bit, linearly-encoded quantizing noise. Such quantizing noise is, in fact, so extreme that the resulting low SNR, in the approximate range of 23 dB as compared with 33–36 dB for multiple eight-bit encoding PCM systems in tandem, is a characteristic and reliable indicator of the presence of ADPCM equipment in the transmission system.

Description of the Invention From further testing with the procedure of the present invention it was determined that the repeatable accuracy is sufficient, i.e. to within about 0.3 dB, to dependably establish the presence of multiple ADPCM devices in tandem in a target system, and to readily compare the individual performance of a number of ADPCM encoders, particularly with respect to their ability to accommodate modems with a frequency spectrum similar to V.29 9600 b/s modems.

PAMS Technology

Patent No. 5,621,854

Title: Method and apparatus for objective speech quality measurements of telecommunication equipment

Date Issued: April 15, 1997

Date of Application: December 12, 1994

Inventors: Hollier; Michael P. (Suffolk, GB2)

Assignee: British Telecommunications public limited company (London, GB2)

Abstract

A telecommunications testing apparatus comprising a signal generator which generates a speech-like synthetic signal, which is supplied to the input of a telecommunication apparatus to be tested. The distorted output of the telecommunications apparatus is supplied to an analyzer, which derives, for both the undistorted test signal and the distorted signal from the telecommunications apparatus, a measure of the excitation of the human auditory system generated by both signals, taking into account both spectral masking and temporal masking phenomena. The difference between the two excitations is then calculated, and a measure of the loudness of the difference is derived which is found to indicate to a high degree of accuracy the human subjective response to the distortion introduced by the telecommunications system.

Citations

U.S. Patent Documents

4860360; Aug., 1989; Boggs; Method of evaluating speech

4972484; Nov., 1990; Theile et al.; Method of transmitting or storing masked sub-band coded audio signals

Other References

John G. Beerends and Jan A. Stemerdink, "Measuring the Quality of Audio Devices," Preprint 3070 (L-8) of a paper presented at the 90th Convention of the Audio Engineering Society, p. 5 Feb. 1991.

Hiroshi Irii, Keita Kurashima, Nobuhiko Kitawaki, and Kenzo Itoh, "Objective Measurement Method for Estimating Speech Quality of Low-Bit-Rate Speech Coding," NTT Review, vol. 3, No. 5, pp. 79–87, Sep. 1991.

Beerends et al, "A Perceptual Audio Quality Measure Based on a Psychoacoustic Sound Representation," J. Audio Eng. Soc., vol. 40, No. 12, 1992, pp. 963–978.

Brandenburg et al, "'NMR' and 'Masking Flag': Evaluation of Quality Using Perceptual Criteria," AES 11th International Conferences, pp. 169–179, 1992.

Kalittsev, "Estimate of the Information Content of Speech Signals Based on Modern Speech Analysis," 1298 Telecommunications and Radio Engineering 47 (1992) Jan., No. 1, New York, US, pp. 11–15.

Zwicker et al, "Audio Engineering and Psychoacoustics: Matching Signals to the Final Receiver, the Human Auditory System," J. Audio Eng. Soc., vol. 39, No. 3, 1991, Mar., pp. 115–126.

Dimolitsas et al, "Objective Speech Distortion Measures and Their Relevance of Speech Quality Assessments," IEE Proceedings, vol. 136, Pt. 1, No. 5, Oct. 1989, pp. 317–324.

Herre et al, "Analysis tool for Realtime Measurements Using Perceptual Criteria," AES 11th International Conference, 1992, pp. 180–190.

Moore et al, "Suggested Formulae for Calculating Auditory-Filter Bandwidths and Excitation Patterns," J. Acoust. Soc. Am. 74(3), Sep. 1983, pp. 750–753.

Extracts

Background of the Invention In testing telecommunications apparatus (for example, a telephone line, a telephone network, or communications apparatus such as a coder) a test signal is introduced to the input of the telecommunications apparatus, and some test is applied to the resulting output of the apparatus. It is known to derive "objective" test measurements, such as the signal to noise ratio, which can be calculated by automatic processing apparatus. It is also known to apply "subjective" tests, in which a human listener listens to the output of the telecommunications apparatus, and gives an opinion as to the quality of the output.

Some elements of telecommunications systems are linear. Accordingly, it is possible to apply simple artificial test signals, such as discrete frequency sine waves, swept sine signals or chirp signals, random or pseudo random noise signals, or impulses. The output signal can then be analyzed using, for example, Fast Fourier Transform (FET) or some other spectral analysis technique. One or more such simple test signals are sufficient to characterise the behaviour of a linear system.

On the other hand, modern telecommunications systems include an increasing number of elements which are nonlinear and/or time variant. For example, modern low bit-rate digital speech coders, forming part of mobile telephone systems, have a nonlinear response and automatic gain controls (AGCs), voice activity detectors (VADs) and associated voice switches, and burst errors contribute time variations to telecommunications systems of which they form [a] part. Accordingly, it is increasingly less possible to use simple test methods developed for linear systems to derive objective [measures] of the distortion or acceptability of telecommunications apparatus.

On the other hand, subjective testing by using human listeners is expensive, time-consuming, difficult to perform, and inconsistent. However, despite these problems the low correlation between objective measures of system performance or distortion and the subjective response of a human user of the system means that such subjective testing remains the best way of testing telecommunications apparatus.

Recently in the paper "Measuring the Quality of Audio Devices" by John G. Beerends and Jan A. Stemerdink, presented at the 90th AES Convention, 1991 Feb. 19–22, Paris, printed in AES Preprints as Preprint 3070 (L-8) by the Audio Engineering Society, it has been proposed to measure the quality of a speech coder for digital mobile radio by using, as test signals, a database of real recorded speech and analyzing the corresponding output of the coder using a perceptual analysis method designed to correspond in some aspects to the processes which are thought to occur in the human ear.

It has also been proposed (for example in "Objective Measurement Method for Estimating Speech Quality of Low Bit Rate Speech Coding," Irii, Kurashima, Kitawaki and Itoh, NTT Review, Vol 3. No. 5, September 1991) to use an artificial voice signal (i.e. a signal which is similar in a spectral sense to the human voice, but which does not convey any intelligence) in conjunction with a conventional distortion analysis measure such as the cepstral distance (CD) measure, to measure the performance of telecommunications apparatus.

It would appear obvious, when testing apparatus such as a coder which is designed to encode human speech, and when employing an analysis method based on the human ear, to use real human speech samples as was proposed in the above paper by Beerends and Stemerdink. In fact, however, the performance of such test systems is not particularly good.

Brief Summary of the Invention Accordingly, it is an object of the invention to provide an improved telecommunications testing apparatus and method. It is another object of the invention to provide a telecommunications testing apparatus which can provide a measure of the performance of [a] telecommunication system which matches the subjective human perception of the performance of the system.

The present invention provides telecommunications testing apparatus comprising a signal generator (7) for supplying a test signal which has a spectral resemblance to human speech but corresponds to more than one speaker, and analysis means (8) for receiving a distorted signal which corresponds to said test signal when distorted by telecommunications apparatus (1) to be tested, and for analyzing said distorted signal to generate a distortion perception measure which indicates the extent to which the distortion of said signal will be perceptible to a human listener.

Patent No. 5,794,188

Title: Speech signal distortion measurement which varies as a function of the distribution of measured distortion over time and frequency

Date Issued: August 11, 1998

Date of Application: April 4, 1996

Inventors: Hollier; Michael Peter (Suffolk, GB)

Assignee: British Telecommunications public limited company (London, GB3)

Abstract

Telecommunications testing apparatus includes an analyzer arranged to receive a distorted signal which corresponds to a test signal when distorted by [the] telecommunications [apparatus] to be tested. The analyzer periodically derives, from the distorted signal, a plurality of spectral component signals responsive to the distortion in each of a plurality of spectral bands, over a succession of time intervals. The analyzer generates a measure of the subjective impact of the distortion due to the telecommunications apparatus, the measure of subjective impact being calculated to depend upon the spread of the distortion over time and/or over the spectral bands.

Citations

U.S. Patent Documents

4860360; Aug., 1989; Boggs; Method of evaluating speech

4972484; Nov., 1990; Theile et al.; Method of transmitting or storing masked sub-band coded audio signals

Other References

Quincy, “Prolog-Based Expert Pattern Recognition System Shell for Technology Independent, User-Oriented Classification of Voice Transmission Quality,” IEEE Int Conf on Communications—Sessions 33.3, vol. 2, 7–10 Jun. 1987, Seattle (US), pp. 1164–1171.

Kubichek et al, “Speech Quality Assessment Using Expert Pattern Recognition Techniques,” IEEE Pacific RIM Conference on Communications, Computers and Signal Processing, 1–2 Jun. 1989, Victoria (CA), pp. 208–211, XP000077468.

Patent Abstracts of Japan, vol. 17, No. 202 (E-1353), 20 Apr. 1993 & JP-A-04 345327 (Nippon Telegr&Teleph Corp), 1 Dec. 1992.

Gierlich, "New Measurement Methods for Determining the Transfer Characteristics of Telephone Terminal Equipment," Proceedings of 1992 IEEE International Symposium on Circuits and Systems, 10–13 May 1992, San Diego (US), New York (US), vol. 4, pp. 2069–2072.

Sobolev, "Estimation of Speech Signal Transmission Quality from Measurements of Its Spectral Dynamics," Telecommunications and Radio Engineering, vol. 47, No. 1, Jan. 1992, Washington US, pp. 16–21, XP000316414.

Beerends, "A Perceptual Audio Quality Measure Based on a Psychoacoustic Sound Representation," J. Audio Eng. Soc., vol. 40, No. 12, 1992, pp. 963–978.

Brandenburg et al, "'NMR' and 'Masking Flag': Evaluation of Quality Using Perceptual Criteria," AES 11th International Conference, pp. 169–179, 1992.

Zwicker et al, "Audio Engineering and Psychoacoustics: Matching Signals to the Final Receiver, the Human Auditory Sytem," J. Audio Eng. Soc., vol. 39, No. 3, 1991, pp. 115–126.

Irii et al, "Objective Measurement Method for Estimating Speech Quality of Low-Bit-Rate Speech Coding," NTT Review, vol. 3, No. 5, Sep. 1991, pp. 79–87.

Dimolitsas, et al, "Objective Speech Distortion Measures and Their Relevance to Speech Quality Assessments," IEE Proceedings, vol. 136, Pt. I, No. 5, Oct. 1989, pp. 317–324.

Herre et al, "Analysis Tool for Realtime Measurements Using Perceptual Criteria," AES 11th International Conference, 1992.

Kalittsev, "Estimates of the Information Content of Speech Signals," 1298 Telecommunications and Radio Engineering 47 (1992), Jan., No. 1, New York, US, pp. 11–15.

Beerends et al, "Measuring the Quality of Audio Devices," AES 90th Convention, (19 Feb. 1991), pp. 1–8.

Moore et al, "Suggested Formulae for Calculating Auditory-Filter Bandwidths and Excitation Patterns," J. Acoust. Soc. Am. 74(3), Sep. 1983, pp. 750–753.

Extensions of PAMS Technology

Patent No. 5,999,900

Title: Reduced redundancy test signal similar to natural speech for supporting data manipulation functions in testing telecommunications equipment

Date Issued: December 7, 1999

Date of Application: June 19, 1998

Inventors: Hollier; Michael Peter (Suffolk, GB)

Assignee: British Telecommunications public limited company (London, GB)

Abstract

A test signal data structure for supporting the data manipulation functions during testing of a telecommunications apparatus includes a succession of segments of real or synthetic speech which includes different successions of the same sounds; rarely occurring sounds; and sounds at extremes of level, pitch and duration, so as to be similar to natural speech with redundancy removed. In analyzing the performance of the telecommunications apparatus the analyzer takes into account the frequency of occurrence of each sound in natural speech.

Citations

U.S. Patent Documents

4218587; Aug., 1980; Elder et al.; Complex signal generation and transmission

4352182; Sep., 1982; Billi et al.; Method of and device for testing the quality of digital speech-transmission equipment

4446341; May, 1984; Rubin; Mechanized testing of subscriber facilities

4449231; May, 1984; Chytil; Test signal generator for simulated speech

4780885; Oct., 1988; Paul et al.; Frequency management system

4860360; Aug., 1989; Boggs; Method of evaluating speech

4972484; Nov., 1990; Theile et al.; Method of transmitting or storing masked sub-band coded audio signals

5195124; Mar., 1993; Ishioka; Testing system for local subscribers

5369644; Nov., 1994; Oliver; Method for testing a communication system using canned speech

5392381; Feb., 1995; Furuya et al.; Acoustic analysis device and a frequency conversion device used therefor

5425076; Jun., 1995; Knippelmier; Cellular communications test system

5448624; Sep., 1995; Hardy et al.; Telephone network performance monitoring method and system

5490234; Feb., 1996; Narayan; Waveform blending technique for text-to-speech system

5621854; Apr., 1997; Hollier; Method and apparatus for objective speech quality measurements of telecommunication equipment

5642113; Jun., 1997; Immink; Methods and devices for converting a sequence of m-bit information words to a modulated signal and including that signal on a record carrier, devices for decoding that signal and reading it from a record carrier, and that signal

Other References

Scott A. Horstemeyer and Daniel J. Santos, "A New Frontier in Patents: Patent Claims to Propagated Signals," Intellectual Property Today, pp. 16–21, Jul. 1998.

Beerends, "A Perceptual Audio Quality Measure Based on a Psychoacoustic Sound Representation," J. Audio Eng. Soc., vol. 40, No. 12, 1992, pp. 963–978.

Brandenburg et al, "'NMR' and 'Masking Flag': Evaluation of Quality Using Perceptual Criteria," AES 11th International Conference, pp. 169–179, 1992.

Zwicker et al, "Audio Engineering and Psychoacoustics: Matching Signals to the Final Receiver, the Human Auditory System," J. Audio Eng. Soc., vol. 39, No. 3, 1991, pp. 115–126.

Irii et al, "Objective Measurement Method for Estimating Speech Quality of Low-Bit-Rate Speech Coding," NTT Review, vol. 3, No. 5, Sep. 1991, pp. 79–87.

Dimolitsas et al, "Objective Speech Distortion Measures and Their Relevance to Speech Quality Assessments," IEE Proceedings, vol. 136, Pt. 1, No. 5, Oct. 1989, pp. 317–324.

Herre et al, "Analysis Tool for Realtime Measurements Using Perceptual Criteria," AES 11.sup.th International Conference, 1992, pp. 180–190.

Kalittsev, "Estimate of the Information Content of Speech Signals," 1298 Telecommunications and Radio Engineering 47 (1992) Jan., No. 1, New York, US, pp. 11–15.

Beerends et al, "Measuring the Quality of Audio Devices," AES 90.sup.th Convention (Feb. 19, 1991), pp. 1–8.

Moore et al, "Suggested Formulae for Calculating Auditory-Filter Bandwidths and Excitation Patterns," J. Acoust. Soc. Am. 74(3), Sep. 1983, pp. 750–753.

Kubichek et al, "Speech Quality Assessment Using Expert Pattern Recognition Techniques," IEEE Pacific Rim Conference on Communications, Computers and Signal Processing, Jun. 1–2, 1989, pp. 208–211.

Quincy, "Prolog-Based Expert Pattern Recognition System Shell for Technology Independent, User-Oriented Classification of Voice Transmission Quality," IEEE, pp. 1164–1171, 1987.

Patent Abstract of Japan, vol. 17, No. 202 (E-1353), Apr. 20, 1993, JP-A-04-345327 (Nippon Telegr&Teleph Corp), Dec. 1, 1992.

Sobolev, "Estimation of Speech Signal Transmission Quality from Measurements of ITS Spectral Dynamics," 1992 Scripta Technica, Inc., pp. 16–21, No. 8, 1991.

Gierlich, "New Measurement Methods for Determining the Transfer Characteristics of Telephone Terminal Equipment," Proceedings of 1992 IEEE International Symposium on Circuits and Systems, May 10–13, 1992, San Diego (US), New York (US), vol. 4, pp. 2069–2072.

Patent No. 5,890,104

Title: Method and apparatus for testing telecommunications equipment using a reduced redundancy test signal

Date Issued: March 30, 1999

Date of Application: December 14, 1995

Inventors: Hollier; Michael Peter (Suffolk, GB)

Assignee: British Telecommunications public limited company (London, GB2)

Abstract

A test signal for a telecommunications apparatus includes a succession of segments of real or synthetic speech which includes different successions of the same sounds; rarely occurring sounds; and sounds at extremes of level, pitch and duration, so as to be similar to natural speech with redundancy removed. In analyzing the performance of the telecommunications apparatus the analyzer takes into account the frequency of occurrence of each sound in natural speech.

Citations

U.S. Patent Documents

4352182; Sep., 1982; Billi et al.; Method of and device for testing the quality of digital speech transmission equipment

4860360; Aug., 1989; Boggs; Method of evaluating speech

4972484; Nov., 1990; Theile et al.; Method of transmitting or storing masked sub-band coded audio signals

5195124; Mar., 1993; Ishioka; Testing system for local subscribers

5369644; Nov., 1994; Oliver; Acoustic analysis device and a frequency conversion device used therefor

5490234; Feb., 1996; Narayan; Waveform blending technique for text-to-speech system

Extracts

Summary of the Invention The present invention is generally based on the realisation that physiological and linguistic constraints prevent certain combinations of speech sounds from occurring in natural speech, and the realisation that for testing purposes there is considerable redundancy in many of the sounds produced in human speech. Accordingly, according to the present invention, speech sounds from real human speech (either recorded or synthesised) are assembled into a structure which provides representative examples of a variety of speech sounds and levels, without redundancy, to provide a speech like test signal of feasible length. In this manner, a speech signal can be obtained which includes the formant structure, and the temporal structure, of natural speech...whilst remaining relatively representative and of realistic duration.

Patent No. 5,799,133

Title: Training process

Date Issued: August 25, 1998

Date of Application: June 27, 1996

Inventors: Hollier; Michael P. (Ipswich, GB2); Gray; Philip (Ipswich, GB2)

Assignee: British Telecommunications public limited company (London, GB2)

Abstract

A training apparatus for establishing the network definition function of a trainable processing apparatus, such as, for example, a neural network, is disclosed. The training apparatus analyzes a signal and includes means for providing a training sequence, the training sequence including a first signal and a distorted version of the first signal. The training sequence is transmitted to an analysis means which generates a distortion perception measure that indicates the extent to which the distortion would be perceptible to a human observer. The network definition function is determined by applying the distortion perception measure to the trainable processing apparatus.

Citations

U.S. Patent Documents

4860360; Aug., 1989; Boggs; Method of evaluating speech

4972484; Nov., 1990; Theile et al.; Method of transmitting or storing masked sub-band coded audio signals

5621854; Apr., 1997; Hollier; Method and apparatus for objective speech quality measurements of telecommunication equipment

5630019; May, 1997; Kochi; Waveform evaluating apparatus using neural network

Other References

Beerends, "A Perceptual Audio Quality Measure Based on a Psychoacoustic Sound Representation," J. Audio Eng. Soc., vol. 40, No. 12, 1992, pp. 963–978.

Brandenburg et al, "'NMR' and 'Masking Flag': Evaluation of Quality Using Perceptual Criteria," AES 11th International Conference, pp. 169–179, 1992.

Zwicker et al, "Audio Engineering and Psychoacoustics: Matching Signals to the Final Receiver, the Human Auditory System," J. Audio Eng. Soc., vol. 39, No. 3, 1991, pp. 115–126.

Irii et al, "Objective Measurement Method for Estimating Speech Quality of Low-Bit-Rate Speech Coding," NTT Review, vol. 3, No. 5, Sep. 1991, pp. 79–87.

Dimolitsas et al, "Objective Speech Distortion Measures and Their Relevance to Speech Quality Assessments," IEE Proceedings, vol. 136, Pt. 1, No. 5, Oct. 1989, pp. 317–324.

Herre et al, "Analysis Tool for Realtime Measurements Using Perceptual Criteria," AES 11th International Conference, 1992.

Kalittsev, "Estimate of the Information Content of Speech Signals," 1298 Telecommunications and Radio Engineering 47 (1992), Jan., No. 1, New York, US, pp. 11–15.

Beerends et al, "Measuring the Quality of Audio Devices," AES 90th Convention, (19 Feb. 1991), pp. 1–8.

Moore et al, "Suggested Formulae for Calculating Auditor-Filter Bandwidths and Excitation Patterns," J. Acoust. Soc. Am. 74(3), Sep. 1983, pp. 750–753.

Gierlich, "New Measurement Methods for Determining the Transfer Characteristics of Telephone Terminal Equipment," Proceedings of 1992 IEEE International Symposium on Circuits and Systems, 10–13 May 1992, San Diego (US), New York (US), vol. 4, pp. 2069–2072.

Sobolev, "Estimates of Speech Signal Transmission Quality from Measurements of Its Spectral Dynamics," Telecommunications and Radio Engineering, vol. 47, No. 1, Jan. 1992, Washington US, pp. 16–21, XP000316414.

Quincy, "Prolog-Based Expert Pattern Recognition System Shell for Technology Independent, User-Oriented Classification of Voice Transmission Quality," IEEE Int Conf on Communications—Sessions 33.3, vol. 2, 7–10 Jun. 1987, Seattle (US), pp. 1164–1171.

Kubichek et al, "Speech Quality Assessment Using Expert Pattern Recognition Techniques," IEEE Pacific Rim Conference on Communications, Computers and Signal Processing, 1–2 Jun. 1989, Victoria (CA), pp. 208–211, XP000077468.

Patent Abstracts of Japan, vol. 17, No. 202 (E-1353), 20 Apr. 1993 & JP-A-04 345 327 (Nippon Telegr&Teleph Corp), 1 Dec. 1992.

Extracts

Background of the Invention Signals carried over telecommunications links can undergo considerable transformations, such as digitisation, data compression, data reduction, amplification, and so on. All of these processes can distort the signals. For example, in digitising a waveform whose amplitude is greater than the maximum digitisation value, the peaks of the waveform will be converted to a flat-topped form (a process known as peak clipping). This adds unwanted harmonics to the signal. Distortions can also be caused by electromagnetic interference from external sources.

Many of the distortions introduced by the processes described above are non-linear, so that a simple test signal may not be distorted in the same way as a complex waveform such as speech, or at all. For a telecommunications link carrying data it is possible to test the link using all possible data characters (e.g. the two characters 1 and 0 for a binary link, or the twelve tone-pairs used in DTMF (dual tone multi-frequency) systems[)]. However speech does not consist of a limited number of well-defined signal elements, but is a continuously varying signal, whose elements vary according not only to the content of the speech (and the language used) but also the physiological and psychological characteristics of the individual speaker, which affect characteristics such as pitch, volume, characteristic vowel sounds etc.

It is known to test telecommunications equipment by running test sequences using samples of speech. Comparison between the test sequence as modified by the equipment under test and the original test sequence can be used to identify distortion introduced by the equipment under test. However, these arrangements require the use of a pre-arranged test sequence, which means they cannot be used on live telecommunications links—that is, links currently in use—because the test sequence would interfere with the traffic being carried and be audible to the users, and because conversely the live traffic itself (whose content cannot be predetermined) would be detected by the test equipment as distortion of the test signal.

In order to carry out tests on equipment in use, without interfering with the signals being carried by the equipment (so-called non-intrusive testing), it is desirable to carry out the tests using the live... speech signals themselves as the test signals. However, a problem with using live speech as the test signal is that there is no instantaneous way of obtaining, at the point of measurement, a sample of the original signal. Any means by which the original signal might be transmitted to the measurement location would be likely to be subject to similar distortions as the link under test....

Summary of the Invention According to a first aspect of the present invention, there is provided a training apparatus for establishing the network definition function of a trainable processing apparatus for analysing a signal, the training apparatus comprising means for providing a training sequence comprising a first signal and a distorted version of the first signal, analysis means for receiving the training sequence and generating a distortion perception measure for indicating the extent to which the distortion would be perceptible to a human observer, and means for applying the distortion perception measure to the trainable processing apparatus to determine the network definition function.

In a further aspect the invention comprises a method of establishing the network definition function of a trainable process for analysing a signal, comprising the steps of providing a training sequence comprising a first signal and a distorted version of the first signal, and determining the network definition function by...measuring the perceptual degree of distortion present in each segment, as determined by an analysis process comprising generating a distortion perception measure which indicates the extent to which the distortion of said signal will be perceptible to a human observer.

In a preferred arrangement the analysis process estimates the effect which would be produced on the human sensory system by distorted and undistorted versions of the same signal, and determines the differences between the said effects, and generates said distortion perception measure in dependence upon said difference.

Patent No. 5,848,384

Title: Analysis of audio quality using speech recognition and synthesis

Date Issued: December 8, 1998

Date of Application: January 13, 1997

Inventors: Hollier; Michael Peter (Ipswich, GB); Sheppard; Philip John (Ipswich, GB)

Assignee: British Telecommunications Public Limited Company (London, GB2)

Abstract

An apparatus for monitoring signal quality in a communications link is provided which recognizes speech elements in signals received over the communications link and generates therefrom an estimate of the original speech signal, and compares the estimated signal with the actual received signal to provide an output based on the comparison.

Citations

U.S. Patent Documents

4063031; Dec., 1977; Grunza; System for channel switching based on speech word versus noise detection

Other References

Wang S, Sekey A, Gersho A, "An Objective Measure for Predicting Subjective Quality of Speech Coders," IEEE J. on Selected Areas in Communications, Vol. 10, No. 5, June 1992.

Beerends J G, Stemerdink J A, "A Perceptual Audio Quality Measure Based on a Psychoacoustic Sound Representation," J Audio Eng. Soc., Vol. 40, No. 12, Dec. 1992.

Stuart J R, "Psychoacoustic Models for Evaluating Errors in Audio Systems," Procs. 10A, Vol. 13, Part 7, November 1991.

Hollier M P, Hawksford M O, Guard D R, "Characterisation of Communications Systems Using a Speech-Like Test Stimulus," J Audio Eng. Soc., Vol. 41, No. 12, Dec. 1993.

Halka U, Heuter U, "A New Approach to Objective Quality-Measures Based on Attribute-Matching," Speech Comms, Early 1992.

NTIA, CCITT SG XII Contribution "Effects of Speech Amplitude Normalization on NTIA Objective Voice Quality Assessment Method," DOC SQ-74.91, Dec. 1991.

Irii H, Kozono J, Kurashima K, "PROMOTE-A System for Estimating Speech Transmission Quality in Telephone Networks," NTT Review, Vol. 3, No. 5, September 1991.

Extracts

Description of Related Art When specifying and monitoring performance of a telecommunications system it is highly desirable to apply a measurement which directly reflects those parameters which will determine likely customer satisfaction. By modelling the human senses, e.g. hearing, it is possible to estimate the subjective performance of real systems, e.g. the subjective audio quality. This approach to measurement, known as perceptual analysis, is an important emerging area.

Perceptual analysis techniques are important because of two main benefits:

(i) The technique can predict the performance of a complex, non-linear system, e.g. low-bit-rate speech coding schemes, which it is not feasible to measure with conventional engineering measurement methods.

(ii) The measurement yields a result in terms of the likely subjective audio quality. This is exactly the information required to optimise system design and operation to provide the required subjective quality.

Existing perceptual analysis methods include both analogous and empirical models....

There is a further class of performance measurement which is of considerable commercial significance, but which cannot be addressed with existing perceptual analysis techniques. This requirement is to assess, non-intrusively, the subjective audio quality of network connections and routes carrying live traffic. Such a measurement would be highly advantageous in providing the following facilities:

(i) The provision of connection quality monitoring, allowing the performance of the network to be assessed,

(ii) The provision of information on problem connections so that remedial action can be taken, and

(iii) The automatic provision of information on the subjective performance of particular connections to the "intelligent network" which will be able to route traffic accordingly. This will allow a more direct optimisation of subjective audio quality than is possible with conventional engineering measurement methods.

Summary of the Invention According to the invention there is provided a method of analysis of the audio quality of a system carrying speech signals, comprising the steps of:

Applying to the output of the communications system a speech recognition process;

Generating thereby an estimate of the original speech signal applied to the input of the communications system;

Comparing the estimated input signal with the output signal; and

Generating a performance estimate based on said comparison.

By reconstructing the input signal the application of comparative analysis techniques to nonintrusive measurement is possible....

In a preferred arrangement, the method comprises the additional step of adapting the macro-properties of the speech elements in the estimated signal to match those in the output signal. The term "macro-properties" is used in this specification to mean the properties of each speech element such as pitch and duration which vary between talkers, as distinct from the microstructure which determines the individual phonemes being spoken.

Patent No. 5,940,792

Title: Nonintrusive testing of telecommunication speech by determining deviations from invariant characteristics or relationships

Date Issued: August 17, 1999

Date of Application: January 13, 1997

Inventors: Hollier; Michael P. (Ipswich, GB)

Assignee: British Telecommunications Public limited company (London, GB)

Abstract

A method of analysis of characteristics of a telecommunications network measures properties of speech carried by a line under test and includes the steps of identifying a part of the speech signal which has a property independent of the talker, and detecting deviations from that property in the received signal, thereby identifying characteristics of the signal imposed by the system. The properties identified may be characteristic waveforms of speech phonemes, in which the fact that the human voice is only capable of making certain sounds is used to determine what deviations from those sounds have been imposed by the system. In an alternative arrangement identifying a part of the speech signal having a property which varies in a predetermined manner in relation to an absolute characteristic of the talker, and deriving the absolute characteristics therefrom [*sic*]. The absolute characteristic may be the absolute level of the voice, and the other property may be a talker-independent function of the absolute level, such as the spectral content of the unvoiced fricatives in the speech.

Citations

U.S. Patent Documents

5313555; May, 1994; Kamiya; Lombard voice recognition method and apparatus for recognizing voices in noisy circumstance

5590242; Dec., 1996; Juang et al.; Signal bias removal for robust telephone speech recognition

5664059; Sep., 1997; Zhao; Self-learning speaker adaptation based on spectral variation source decomposition

5778336; Jul., 1998; Chou et al.; Speech coding and joint data/channel bias estimation using finite state vector quantizer derived from sequential constraints

5794192; Aug., 1998; Zhao; Self-learning speaker adaptation based on spectral bias source decomposition, using very short calibration speech

5812972; Sep., 1998; Juang et al.; Adaptive decision directed speech recognition bias equalization method and apparatus

Other References

Yunxin Zhao, "A New Speaker Adaptation Technique Using Very Short Calibration Speech," Proc. ICASSP 93, vol. II, pp. 562–565, Apr. 1993.

Mazin G. Rahim et al., "Signal Bias Removal for Robust Telephone Based Speech Recognition in Adverse Environments," Proc. ICASSP 94, vol. I, pp. 445–448, Apr. 1994.

Yunxin Zhao, "An Acoustic-Phonetic-Based Speaker Adaptation Technique for Improving Speaker-Independent Continuous Speech Recognition," IEEE Trans. on Speech and Audio Processing, vol. 2, No. 2, pp. 380–394, Jul. 1994.

David B. Ramsden, "In-Service Nonintrusive Measurement on Speech Signals," Proc. GLOBECOM 91, pp. 1761–1764, Dec. 1991.

Extracts

Background of the Invention Modern telecommunications systems perform complex operations on the signals they handle in the process of transmitting the signals through the telecommunications network, for example digitisation and compression techniques. These operations have non-linear effects on the signal inputs and it is thus not possible to model the effects of the network by the simple additive effect of each component of the network. In particular, the effect of the network on speech is not easily derivable from studying its effect on a simple test signal such as a sine wave.

Various methods of deriving test signals which mimic generalised speech properties have been devised...but these must all presuppose certain conditions, and in particular they require the use of predetermined test signals. The use of live (real time) traffic as a test signal for these tests would be impossible. The test site (which may be many thousands of miles away from the signal source in the case of an intercontinental link) needs to have knowledge of the test signal, so that deviations from the test signal can be distinguished from the test signal itself. The use of prearranged test signals may also require cooperation between the operators of two or more networks. Moreover, any line carrying a voice-frequency test signal is not available for use by a revenue-earning call, as the revenue-earning call would interfere with the test, and the test signal would be audible to the makers of the revenue-earning call.

It is known to test lines carrying live data (as distinct from speech), but this is a relatively simple problem because the information content of the signal consists of only a limited range of signals (e.g. DTMF tones, or binary digits), and it is relatively easy to identify elements of the signal which depart from this permitted set. In such arrangements, reliance is placed on the known forms of the permitted signals.

Summary of the Invention The present invention seeks to provide a method of testing a line whilst in use for carrying live speech. A testing system is known in which the signal-to-noise ratio, or other measurable characteristics, of the system are determined by classifying samples as speech or as noise and comparing the properties of each sample. This is disclosed in a paper by David B Ramsden in IEEE “Globecom 91” pages 1761 to 1764, and in European patent 0565424. However, this does not attempt to measure the properties of the speech content itself.…

The invention makes use of the fact that although the live speech signal generated at the signal source is not known at the test location, certain characteristics of the signal are known because they are constrained by the fact that the signal is speech and will therefore have certain characteristics peculiar to speech. The invention makes use of this fact by identifying the behaviour of the received signal in relation to these characteristics. Particular classes of properties which may be identified include:

1. *Pseudo-deterministic.* Different talkers use different vowel sounds because of linguistic differences, but these all fall within a small, well-defined group because the human larynx and vocal tract are only capable of producing a limited range of vowel sounds, whose spectral structure is consistent across all talkers. Analysis of the actual spectral content of the vowels in a signal can identify distortions introduced by…the telecommunications system.

2. *Consistently varying characteristics.* Certain properties of speech vary in relation to certain other properties in a consistent way. If one of the properties is measurable at the test location, the value of the other property can be derived from it, even though it is not directly measurable. An example of such a relationship is the spectral variation of voiced fricatives according to the absolute loudness of the speaker's voice. The fricatives are those sounds created when the airstream is forced between two closely spaced articulators....The spectral contents of the fricative sounds vary with the loudness (volume) with which the talker is speaking, and this variation is consistent across the population of talkers. This spectral content can, therefore, indicate the absolute level at which the talker is speaking. The absolute vocal level estimated in this way can be compared with the received signal strength to calculate losses in the telecommunications system. The vocal level of the speaker estimated in this way may also be a useful indicator of signal quality on the return path, as perceived by the speaker, as a person hearing a faint signal will tend to speak louder.

3. *Gross characteristics.* A number of features of conversational speech can be used to identify difficulties the talkers may have in understanding each other. For example, if the talkers are not switching between each other smoothly, but are talking over each other, this can indicate difficulties in hearing each other or confusion over whose turn it is. If several calls on a given route are unusually short, this can also indicate a faulty line, as users are sticking to essential points of the call, or are giving up altogether and terminating the call, possibly to redial in the hope of getting a clearer line on the second attempt.

None of these classes of characteristics are completely invariant between talkers, but vary within known statistical distributions. More reliable measures of the properties of the network can be obtained by measuring a number of the characteristics referred to above, and/or a number of different talkers using the same line on different calls.

4. *Known non-speech signals.* A line may also be monitored for certain types of signals having characteristic sounds, which should not be found accompanying a speech signal, e.g. feedback howl or data signals from a crossed line.

Patent No. 6,035,270

Title: Trained artificial neural networks using an imperfect vocal tract model for assessment of speech signal quality

Date Issued: March 7, 2000

Date of Application: February 3, 1998

Inventors: Hollier; Michael P (Ipswich, GB); Sheppard; Philip J (London, GB); Gray; Philip (Ipswich, GB)

Assignee: British Telecommunications public limited company (London, GB)

Abstract

A speech signal is subjected [to the imperfect] vocal tract analysis model and the output therefrom is analyzed by a neural network. The output from the neural network is compared with the parameters stored in the network definition function, to derive measurement of the quality of the speech signal supplied to the source. The network definition function is determined by applying to the trainable processing apparatus a distortion perception measure indicative of the extent to which a distortion would be perceptible to a human listener.

Citations

U.S. Patent Documents

5715372; Feb., 1998; Meyers et al.; Method and apparatus for characterizing an input signal

5799133; Aug., 1998; Hollier et al.; Training process

5867813; Feb., 1999; Pietro et al.; Method and apparatus for automatically and reproducibly rating the transmission quality of a speech transmission system

Other References

IEEE International Conference on Communications 87—Session 33.3, vol. 2, Jun. 7–10, 1987, Seattle (US), pp. 1164–1171, Edmund A. Quincy, "Prolog Based Expert Pattern Recognition System Shell for Technology…"

Patent Abstracts of Japan, vol. 17, No. 202 (E-2353), Apr. 20, 1993 & JP,A, 04 345327 (Nippon Telegr & Teleph Corp), Dec. 1, 1992.

IEEE Pacific RIM Conference on Communications, Computers and Signal Processing, Jun. 1–2, 1989, Victoria (CA), pp. 208–211, R.F. Kubichek et al, "Speech Quality Assessment Using Expert Pattern Recognition Techniques."

K. K. Paliwal and B. S. Atal, "Efficient Vector Quantization of LPC Parameters at 24 Bits/Frame," Proc. IEEE ICASSP, pp. 661–664, Apr. 1991.

Patent No. 6,119,083

Title: Training process for the classification of a perceptual signal

Date Issued: September 12, 2000

Date of Application: March 19, 1998

Inventors: Hollier; Michael P (Suffolk, GB); Gray; Philip (Suffolk, GB)

Assignee: British Telecommunications public limited company (London, GB)

Abstract

Training apparatus and method for establishing the network definition function of a trainable processing apparatus for analyzing a signal, includes providing a training sequence having a first signal and a distorted version of the first signal, receiving the training sequence and generating a distortion perception measure for indicating the extent to which the distortion would be perceptible to a human observer, and applying the distortion perception measure to the trainable processing apparatus to determine the network definition function.

Citations

U.S. Patent Documents

4860360; Aug., 1989; Boggs; Method of evaluating speech

4972484; Nov., 1990; Theile et al.; Method of transmitting or storing masked sub-band coded audio signals

5301019; Apr., 1994; Citta; Data compression system having perceptually weighted motion vectors

5621854; Apr., 1997; Hollier; Method and apparatus for objective speech quality measurements of telecommunication equipment

5630019; May, 1997; Kochi; Waveform evaluating apparatus using neural network

Other References

Yogeshwar et al. (A New Perceptual Model for Video) Rutgers University, NJ., pp. 188–193, 1990.

Bellini et al. (Analog Fuzzy Implementation of a Perceptual Classifier for Videophone Sequences) Universita di Bologna, Italy, pp. 787–794, Jul. 1996.

IEEE Int Conf on Communications—Session 33.3, vol. 2, Jun. 7–10, 1987, Seattle, US, pp. 1164–1171, Quincy, “Prolog-Based Expert Pattern Recognition System Shell for Technology Independent, User-Oriented Classification of Voice Transmission Quality.”

IEEE Pacific Rim Conference on Communications, Computers and Signal Processing, Jun. 1–2, 1989, Victoria, CA, Kuichek et al, “Speech Quality Assessment Using Expert Pattern Recognition Techniques.”

Patent Abstracts of Japan, vol. 17, No. 202 (E-1353), Apr. 20, 1993 & JP-A-04 345327 (Nippon Telegr&Teleph Corp), Dec. 1, 1992.

Beerends, “A Perceptual Audio Quality Measure Based on a Psychoacoustic Sound Representation,” A. Audio Eng. Soc., vol. 40, No. 12, 1992, pp. 963–978.

Brandenburg et al, “‘NMR’ and ‘Masking Flag’: Evaluation of Quality Using Perceptual Criteria,” AES 11th International Conferences, pp. 169–179, 1992.

Zwicker et al, “Audio Engineering and Psychoacoustics: Matching Signals to the Final Receiver, the Human Auditory System,” J. Audio Eng. Soc., vol. 39, No. 3, 1991, pp. 115–126.

Irii et al, “Objective Measurement Method for Estimating Speech Quality of Low-Bit-Rate Speech Coding,” NTT Review, vol. 3, No. 5, Sep. 1991, pp. 79–87.

Dimolitsas et al, “Objective Speech Distortion Measures and Their Relevance to Speech Quality Assessments,” IEE Proceedings, vol. 136, Pt. 1, No. 5, Oct. 1989, pp. 317–324.

Herre et al, “Analysis Tool for Realtime Measurements Using Perpetual Criteria,” AES 11.sup.th International Conference, 1992.

Kalittsev, “Estimate of the Information Content of Speech Signals,” 1298 Telecommunications and Radio Engineering 47 (1992), Jan., No. 1, New York, US, pp. 11–15.

Moore et al, “Suggested Formulae for Calculating Auditory-Filter Bandwidths and Excitation Patterns,” J. Acoust. Soc. Am, 74 (3), Sep. 1983, pp. 750–753.

Gierlich, “New Measurement Methods for Determining the Transfer Characteristics of Telephone Terminal Equipment,” Proceedings of

1992, IEEE International Symposium on Circuits and Systems, May 10–13, 1992, San Diego (US), New York (US), vol. 4, pp. 2069–2072.

Sobolev, "Estimation of Speech Signal Transmission Quality from Measurements of Its Spectral Dynamics," Telecommunications and Radio Engineering, vol. 47, No. 1, Jan. 1992, Washington, US, pp. 16–21, XP000316414.

PSQM Technology

Patent No. 5,588,089*

Title: Bark amplitude component coder for a sampled analog signal and decoder for the coded signal

Date Issued: December 24, 1996

Date of Application: May 9, 1995

Inventors: Beerends; John G. (The Hague, NL); Muller; Frank (Delft, NL); van Ravesteijn; Robertus L. A. (Voorburg, NL)

Assignee: Koninklijke PTT Nederland N.V. (Groningen, NL)

Abstract

In a speech coder, a sampled analog signal is filtered by a short-term prediction filter. The result, a segmented residual signal, is transformed from a time domain to a frequency domain into several frequency components, each having a frequency-component amplitude. If a number of new amplitudes is calculated by combining the several frequency-component amplitudes, such that the number of new amplitudes is smaller than the several frequency-component amplitudes, a more efficient coder is created. The reduction of the quality of speech coding, due to loss of information, could be decreased if this calculation is based on the so-called Bark scale (critical frequency bands). In a corresponding speech decoder, at the hand of the number of new amplitudes several new frequency-component [amplitudes] are calculated (the number of new amplitudes being smaller than the several new frequency-component amplitudes), which then are inverse transformed from a frequency domain to a time domain into new subsegments. These new subsegments are inverse filtered by an inverse short-term prediction filter to generate a signal which is representative for a sample analog signal.

Citations

U.S. Patent Documents

4742550; May, 1988; Fette; 4800 BPS interoperable relp system

---

*The Bark encoding is the basis for the psychoacoustic processing implemented in PSQM.

4964166; Oct., 1990; Wilson; Adaptive transform coder having minimal bit allocation processing

4991213; Feb., 1991; Wilson; Speech specific adaptive transform coder

5012517; Apr., 1991; Wilson et al.; Adaptive transform coder having long term predictor

5042069; Aug., 1991; Chhatwal; Methods and apparatus for reconstructing non-quantized adaptively transformed voice signals;

Other References

L. R. Rabiner et al, Chapter 8, *Digital Processing of Speech Signals,* Prentice Hall, New Jersey.

Vary et al, Frequenz, vol. 42, No. 2-3, 1988; pp. 85–93, Sprachcodec Fur Dass Europaische Funkfernsprechnetz.

B. Scharf et al, *Handbook of Perception and Human Performance,* Chapter 14, pp. 1–43, Wiley, New York, 1986.

P. Chang et al, "Fourier Transform Vectors Quantisation for Speech Coding," IEEE Transactions and Communications, vol. Com. 35, No. 10, pp. 1059–1068.

Hermansky et al, "Perceptually Based Linear Predictive Analysis of Speech," Mar., 1985, pp. 509–512, vol. 2 of 4, ICASSP '85 IEEE.

Mazor et al, "Adaptive Subbands Excited Transform (ASET) Coding," Apr., 1986, pp. 3075–3078, vol. 4 of 4, ICASSP '86, IEEE.

Yatsuzuka et al, "Hardware Implementation of 9.6/16 KBIT/S APC/MLC Speech Codec and Its Applications for Mobile Satellite Communications," Jun., 1987, pp. 418–424 CC-87, IEEE Conference '87 Seattle.

Fette et al, "Experiments with a High Quality, Low Complexity 4800 bps Residual Excited LPC (RELP) Vocoder," Apr. 1988, pp. 263–266, vol. 1, ICASSP '88, IEEE.

Schroeder et al, "Optimizing Digital Speech Coders by Exploiting Masking Properties, of the Human Ear," Journal Acoustic Soc. of America, Dec., 1979, pp. 1647–1652.

Johnston, "Transform Coding of Audio Signals Using Perceptual Noise Criteria," IEEE Journal on Selected Areas of Communication, vol. 6, No. 2, Feb. 1988.

Atal, "Predictive Coding of Speech at Low Bit Rates," IEEE, Transactions on Communications, vol. 30, No. 4, Apr., 1982.

Patent No. 5,687,281

Title: Bark amplitude component coder for a sampled analog signal and decoder for the coded signal

Date Issued: November 11, 1997

Date of Application: April 28, 1993

Inventors: Beerends; John Gerard (The Hague, NL); Muller; Frank (Delft, NL); van Ravesteijn; Robertus Lambertus Adrianus (Voorburg, NL)

Assignee: Koninklijke PTT Nederland N.V. (Groningen, NL)

Abstract

A sampled analog signal is filtered by a short-term prediction filter. The result, a segmented residual signal, is transformed from a time domain to a frequency domain into several frequency components, each having a frequency-component amplitude. If a number of new amplitudes is calculated by combining the several frequency-component amplitudes, such that the number of new amplitudes is smaller than the several frequency-component amplitudes, a more efficient coder is created. The reduction of the quality of speech coding, due to loss of information, could be decreased if this calculation is based on the so-called Bark scale (critical frequency bands). In a corresponding speech decoder, at the hand of the number of new amplitudes several new frequency-component [amplitudes] are calculated (the number of new amplitudes being smaller than the several new frequency-component amplitudes), which then are inverse transformed from a frequency domain to a time domain into new subsegments. These new subsegments are inverse filtered by an inverse short-term prediction filter to generate a signal which is representative for a sample analog signal.

Citations

U.S. Patent Documents

4742550; May; 1988; Fette; 4800 BPS interoperable relp system

4964166; Oct., 1990; Wilson; Adaptive transform coder having minimal bit allocation processing

4991213; Feb., 1991; Wilson; Speech specific adaptive transform coder

5012517; Apr., 1991; Wilson et al.; Adaptive transform coder having long term predictor

5042069; Aug., 1991; Chhatwal; Methods and apparatus for reconstructing non-quantized adaptively transformed voice signals

Other References

Hermansky et al, "Perceptually Based Linear Predictive Analysis of Speech," Mar., 1985, pp. 509–512, vol. 2 of 4, ICASSP '85 IEEE.

Mazor et al, "Adaptive Subbands Excited Transform (ASET) Coding," Apr., 1986, pp. 3075–3078, vol. 4 of 4, ICASSP '86, IEEE.

Yatsuzuka et al, "Hardware Implementation of 9.6/16 KBIT/S APC/MLC Speech Codec and its Applications for Mobile Satellite Communications," Jun., 1987, pp. 418–424 CC-87, IEEE Conference '87 Seattle.

Fette et al, "Experiments with a High Quality, Low Complexity 4800 bps Residual Excited LPC (RELP) Vocoder," Apr. 1988, pp. 263–266, vol. 1, ICASSP '88, IEEE.

Schroeder et al, "Optimizing Digital Speech Coders by Exploiting Masking Properties of the Human Ear," Journal Acoustic Soc. of America, Dec., 1979, pp. 1647–1652.

Johnston, "Transform Coding of Audio Signals Using Perceptual Noise Criteria," IEEE Journal on Selected Areas of Communication, Vol. 6, No. 2, Feb., 1988.

Atal, "Predictive Coding of Speech at Low Bit Rates," IEEE, Transactions on Communications, vol. 30, No. 4, Apr., 1982.

L.R. Rabiner et al, Chapter 8, *Digital Processing of Speech Signals,* Prentice Hall, New Jersey, pp. 396–461.

Vary et al, Frequenz, vol. 42, No. 2-3, 1988; pp. 85–93, Sprachcodec Fur Dass Europaische Funkfernsprechnetz.

B. Scharf et al, *Handbook of Perception and Human Performance,* Chapter 14, pp. 1–43, Wiley, New York, 1986.

P. Chang et al, "Fourier Transform Vectors Quantisation for Speech Coding," IEEE Transactions and Communications, vol. Com. 35, No. 10, pp. 1059–1068.

Beranek, *Acoustics,* McGraw-Hill Book Company, Inc., 1954, pp. 332–334.

Patent No. 6,041,294

Title: Signal quality determining device and method

Date Issued: March 21, 2000

Date of Application: September 5, 1997

Inventors: Beerends; John Gerard (The Hague, NL)

Assignee: Koninklijke PTT Nederland N.V. (NL)

Abstract

A device for determining the quality of an output signal to be generated by a signal processing circuit with respect to a reference signal. The device is provided with a first series circuit for receiving the output signal and a second series circuit for receiving the reference signal. The device generates an objective quality signal through a combining circuit which is coupled to the two series circuits. Poor correlation between the objective quality and subjective quality signals, the latter which will be assessed by human observers, can be considerably improved by a differential arrangement present in the combining circuit. This arrangement determines a difference between the two series circuit signals and reduces this difference by a certain value, preferably one that is a function of a series circuit signal. Poor correlation can be improved further by disposing a scaling circuit, between the two series circuits, for scaling at least one series circuit signal. Furthermore, the quality signal can also be scaled as a function of the scaling circuit.

Citations

U.S. Patent Documents

4860360; Aug., 1989; Boggs; Method of evaluating speech

5588089; Dec., 1996; Beerends et al.; Bark amplitude component coder for a sampled analog signal and decoder for the coded signal

5687281; Nov., 1997; Beerends et al.; Bark amplitude component coder for a sampled analog signal and decoder for the coded signal

Other References

Beerends, et al, "A Perceptual Speech-Quality Measure Based on a Psychoacoustic Sound Representation," Journal of the Audio Engineering Society, vol. 42, No. 3, Mar. 1994, pp. 115–123.

Beerends, et al, "A Perceptual Audio Quality Measure Based on a Psychoacoustic Sound Representation," Journal of the Audio Engineering Society, vol. 40, No. 12, Dec. 1992, pp. 963–978.

Beerends, et al, "Modelling a Cognitive Aspect in the Measurement of the Quality of Music Codes," An Audio Engineering Society Preprint, presented at the 96.sup.th Convention, Feb. 26–Mar. 1, 1994, pp. 1–13.

Extracts

Summary of the Invention The object of the invention is, inter alia, to provide a device in which the objective quality signal which is to be assessed by means of the device, and a subjective quality signal, which is to be assessed by human observers have an improved correlation with each other.

For this purpose, the device according to the invention has the characteristic that the differential arrangement is provided with an adjusting arrangement, for reducing the amplitude of the differential signal.

The invention is based, inter alia, on the insight that the poor correlation between objective quality signals, to be assessed by means of known devices, and subjective quality signals, to be assessed by human observers, is the consequence, inter alia, of the fact that certain distortions are found to be more objectionable by human observers than other distortions. This poor correlation is improved by using the two compressing arrangements, and is furthermore based, inter alia, on the insight that the two compressing arrangements do not function optimally, as a consequence of which the amplitude of the differential signal can be reduced, i.e. adjusted, for example by subtracting a signal having a constant value.

The problem of the poor correlation is thus solved by providing the differential arrangement with the adjusting arrangement.

Patent No. 6,064,946

Title: Signal quality determining device and method

Date Issued: May 16, 2000

Date of Application: September 5, 1997

Inventors: Beerends; John Gerard (The Hague, NL)

Assignee: Koninklijke PTT Nederland N.V. (NL)

Abstract

A device for determining the quality of an output signal to be generated by a signal processing circuit with respect to a reference signal is provided with a first series circuit for receiving the output signal and with a second series circuit for receiving the reference signal. The device generates an objective quality signal by means of a combining circuit coupled to the two series circuits. Poor correlation between the objective quality signal and a subjective quality signal, to be assessed by human observers, can be considerably improved by disposing a scaling circuit between the two series circuits for scaling at least one series circuit signal. Furthermore, it is also possible to scale the quality signal as a function of the scaling circuit. Poor correlation can be further improved by determining, using a differential arrangement present in the combining circuit, a difference between the two series circuit signals, and then modifying the difference by a certain value, preferably as a function of a series circuit signal.

Citations

U.S. Patent Documents

4860360; Aug., 1989; Boggs; Method of evaluating speech

5602961; Feb., 1997; Kolesnik et al.; Method and apparatus for speech compression using multi-mode code excited linear predictive coding

Other References

Beerends, et al, "A Perceptual Speech-Quality Measure Based on a Psychoacoustic Sound Representation," Journal of the Audio Engineering Society, vol. 42, No. 3, Mar. 1994, pp. 115–123.

Beerends, et al, "A Perceptual Audio Quality Measure Based on a Psychoacoustic Sound Representation," Journal of the Audio Engineering Society, vol. 40, No. 12, Dec. 1992, pp. 963–978.

Beerends, et al, "Modelling a Cognitive Aspect in the Measurement of the Quality of Music Codes," An Audio Engineering Society Preprint, presented at the 96.sup.th Convention, Feb. 26–Mar. 1, 1994, pp. 1–13.

John G. Beerends and Jan A. Stemerdink, "A Perceptual Audio Quality Measure Based on a Psychoacoustic Sound Representation, Journal of the Audio Engineering Society," pp. 963–978, Dec. 1992.

Extracts

Summary of the Invention The object of the invention is, inter alia, to provide a device in which the objective quality signal to be assessed by means of the device and a subjective quality signal to be assessed by human observers have an improved correlation with each other.

For this purpose, the device according to the invention has the characteristic that the device comprises a scaling circuit which is situated between the first series circuit and the second series circuit, which scaling circuit is provided with a further integrating arrangement for integrating a first series circuit signal and a second series circuit signal with respect to frequency, and a comparing arrangement, coupled to the further integrating arrangement, for comparing the two integrated series circuit signals and for scaling at least one series circuit signal in response to the comparison.

...the two series circuit signals are integrated with respect to frequency and then compared, after which at least one series circuit signal is scaled in response to the comparison...

Due to this scaling, good correlation is obtained between the objective quality signal to be assessed by means of the device, and a subjective quality signal to be assessed by human observers.

Technology Employed in the VQES

Patent No. 6,246,978

Title: Method and system for measurement of speech distortion from samples of telephonic voice signals

Date Issued: June 12, 2001

Date of Application: May 18, 1999

Inventors: Hardy; William C. (Dallas, TX)

Assignee: MCI WorldCom, Inc. (Washington, DC)

Abstract

A system that provides measurements of speech distortion that correspond closely to user perceptions of speech distortion is provided. The system calculates and analyzes first and second discrete derivatives to detect and determine the incidence of change in the voice waveform that would not have been made by human articulation because natural voice signals change at a limited rate. Statistical analysis is performed of both the first and second discrete derivatives to detect speech distortion by looking at the distribution of the signals. For example, the kurtosis of the signals is analyzed as well as the number of times these values exceed a predetermined threshold. Additionally, the number of times the first derivative data is less than a predetermined low value is analyzed to provide a level of speech distortion and clipping of the signal due to lost data packets.

Citations

U.S. Patent Documents

5307441; Apr., 1994; Tzeng; Wear-toll quality 4.8 kbps speech codec

5448624; Sep., 1995; Hardy et al.; Telephone network performance monitoring method and system

5450522; Sep., 1995; Hermansky et al.; Auditory model for parametrization of speech

5682463; Oct., 1997; Allen et al.; Perceptual audio compression based on loudness uncertainty

5699479; Dec., 1997; Allen et al.; Tonality for perceptual audio compression based on loudness uncertainty

5778335; Jul., 1998; Ubale et al.; Method and apparatus for efficient multiband celp wideband speech and music coding and decoding

5943647; Aug., 1999; Ranta; Speech recognition based on HMMs

Extracts

Background of the Invention Various techniques have been used in an attempt to measure speech distortion in digitally mastered waveforms and pseudo speech signals to predict user perception of speech distortion under various conditions. For example, a technique known as PAMS, that was developed in the United Kingdom, uses a recording of digitally mastered phonemes. According to this process, the digitally mastered phonemes are transmitted over a telephone system and recorded at the receiving end. The recorded signal is processed and compared to the originally transmitted signal to provide a measurement of the level of distortion of the transmitted signal....

Further, each of these techniques [is] only effective when known signals are transmitted. The PAMS technique requires the transmission of a special signal containing special phonemes and a comparison of the transmitted signal with the received signal. The second technique requires transmission of sinusoidal waveforms on the audio channel. It would therefore be advantageous to provide a system that would allow measurement and interpretation of speech distortion that uses samples of natural speech from live telephone conversations and does not require the introduction of special signals or comparison with an original signal. It would also be advantageous to be able to sample such signals in a nonintrusive monitoring situation that enables collection of data from live conversations.

Summary of the Invention The present invention overcomes the disadvantages and limitations of the prior art by providing an apparatus and method that allows non-intrusive sampling of live telephone calls and processing of data from those calls to provide a measurement of the level of speech distortion of voice signals.

Patent No. 6,370,120

Title: Method and system for evaluating the quality of packet-switched voice signals

Date Issued: April 9, 2002

Date of Application: December 24, 1998

Inventors: Hardy; William Christopher (Dallas, TX)

Assignee: MCI WorldCom, Inc. (Washington, DC)

Abstract

Method and apparatus for evaluating the quality of a packet-switched voice connection. The apparatus includes measurement circuitry coupled to receive a voice signal. The measurement circuitry measures rate of packet loss and at least one other objective characteristic of the voice signal. The rate of packet loss and the at least one other objective characteristic are related to a plurality of quality characteristics affecting the quality of the voice signal as perceived by users, such that measurements of the rate of packet loss and the at least one other objective characteristic can be translated into subjective quantifications of each of the quality characteristics. A database stores an effects matrix. The effects matrix provides quality information for voice signals that include different combinations of subjective quantifications for each quality characteristic. Software operating on the apparatus utilizes measurements of the rate of packet loss and the at least one other objective characteristic and the effects matrix to generate quality information for the voice signal.

Citations

U.S. Patent Documents

5119367; Jun., 1992; Kawakatsu et al.; Method and a node circuit for routing bursty data

5200950; Apr., 1993; Foglar et al.; Method and circuit arrangement for reducing the loss of message packets that are transmitted via a packet switching equipment

5553059; Sep., 1996; Emerson et al.; Network interface unit remote test pattern generation

5825849; Oct., 1998; Garland et al.; Loop-back test system using a suppressed ringing connection

6041048; Mar., 2000; Erickson et al.; Method for providing information packets from a packet switching network to a base site and corresponding communication system

6046979; Apr., 2000; Bauman; Method and apparatus for controlling the flow of variable-length packets through a multiport switch

6067350; May, 2000; Gordon; Long distance telephone communication system and method

6111949; Aug., 2000; Sheets et al.; Method of rapid automatic hybrid balancing

6128291; Oct., 2000; Perlman et al.; System and method for establishing a call telecommunications path

6147988; Nov., 2000; Bartholmew et al.; IP packet switching in a Telco switch

6243373; Jun., 2001; Turcok; Method and apparatus for implementing a computer network/internet telephone system

6275797; Aug., 2001; Randic; Method and apparatus for measuring voice path quality by means of speech recognition

6282192; Aug., 2001; Murphy et al.; PSTN fallback using dial on demand routing scheme

Extracts

Summary In one aspect, the present invention provides a technique for assessing the quality of a packet-switched communications channel. For instance, the technique provides a means by which a set of objective measurements for a packet-switched telephony connection can be processed to derive a projected perceptual quality level for the connection. It extends upon the techniques of the prior art to adapt for the effects of important new technologies.

Specifically, the present invention addresses packet-switched (e.g., Internet Protocol based) telephony, which is subject to a different set of impairments from ordinary link-switched telephony.

REFERENCES

1. Hardy, William C., *QoS—Measurement and Evaluation of Telecommunications Quality of Service,* John Wiley & Sons, Ltd., Chichester, United Kingdom, 2001.
2. Davidson, Jonathan, and Peters, James, *Voice over IP Fundamentals* Cisco Press, Indianapolis, 2000.
3. ITU-T Recommendation P.800, Methods for subjective determination of transmission quality. Geneva 08/1996.
4. ITU-T Recommendation P.861, Objective quality measurement of telephone-band (300–3400 Hz) speech codecs. Geneva 02/1998.
5. ITU-T Recommendation P.862, Perceptual evaluation of speech quality (PESQ), an objective method for end-to-end speech quality assessment of narrowband telephone networks and speech codecs. Geneva 02/2001.
6. Cavanaugh, J. R., Hatch, R. W., and Sullivan J. L., "Transmission Rating Models for Use in Planning of Telephone Networks," GLOBECOM '83 Conference Record, Vol. 2, IEEE 83CH1956-2, pp. 683–688.
7. ITU-T Recommendation G.107, The E-model, a computational model for use in transmission planning. Geneva 05/2000.
8. ITU-T Recommendation P.561, In-Service, Non-Intrusive Measurement Device—Voice Service Measurements, Geneva 02/1996.
9. ITU-T Recommendation P.562, Analysis and interpretation of INMD voice-service measurements, Geneva, 05/2000.
10. Douskalis, Bill, *Putting VoIP to Work—Softswitch Network Design and Testing,* Prentice Hall, Upper Saddle River, N.J., 2002.

INDEX

A

B

D

G

H

I

J

K

L

M

N

Q

R

S

T

U

V

W

About the Author

Dr. William C. ("Chris") Hardy is the erstwhile principal analyst for quality measurement and analyses at WorldCom. He has a Ph.D. in mathematics and nearly 35 years' experience in operations analysis of communications and computer-based decision support systems gained in the course of: a 13-year career in analyses of military communications systems begun when he joined the Center for Naval Analyses in 1967; 2½ years at Satellite Business Systems as Assistant Director for Systems Measurements; and more than 13 years at MCI/WorldCom, where his principal role was to serve as corporate "fair witness" for service quality, producing, where necessary, operationally meaningful, scientifically defensible analyses of the quality of WorldCom services and their relative position in the highly competitive telecommunications marketplace. He holds 12 patents for telecommunications test technology and has 9 other active patent applications for innovations in test and analysis of commercial telecommunications services.

Dr. Hardy has authored a 1975 treatise for the U.S. Navy entitled "Operational Test and Evaluation for Communications Systems"; and *QoS: Measurement and Evaluation of Telecommunications Service Quality,* released in June 2001 by John Wiley and Sons. His column "Telecom Tips and Quality Quandaries" appeared regularly in *QSDG Magazine,* the official magazine of the Quality of Service Development Group of the ITU, while it was published, and he continues the discourse in the electronic version of the magazine. His other publications include three articles in academic journals, including two in the *Proceedings of the Royal Society of Edinburgh,* and a paper in *GLOBECOM '85* entitled "Subjective Voice Quality Evaluation in a Satellite Communications Environment."

He is an honorary member of the Russian Academy of Science.